CONSTRUCTION ESTIMATING

A Step-by-Step Guide to a Successful Estimate

Karl F. Schmid, PE; LEED AP

 MOMENTUM PRESS

Momentum Press, LLC, New York

Published by Momentum Press®, LLC
222 East 46th Street, New York, NY 10017
www.momentumpress.net

ISBN-13: 978-1-60650-292-1 (softcover)
ISBN-10: 1-60650-292-1 (softcover)
ISBN-13: 978-1-60650-293-8 (e-book)
ISBN-10: 1-60650-293-X (e-book)

Cover Design by Jonathan Pennell
Interior Design by Scribe, Inc. (www.scribenet.com)

10 9 8 7 6 5 4 3 2 1

Printed in the United States of America

To the small contractor, who must keep employees working, the company viable, and the lowest bid–priced jobs on time and within budget

CONTENTS

INTRODUCTION

Go online and type in "construction, estimating" and you will get over 25 estimating and takeoff software programs. There are programs for all the trades and for large, midsized, and small contractors, in both building and heavy construction. Now type in "construction contractor, failures" and scan the various responses. You will learn that according to the Surety Information Office the contractor failure rate is about 25% for trade contractors and 22% for heavy, highway contractors.[1] These figures have been fairly constant for years. You will also learn that poor estimating is a major cause of these failures.

This book will focus on small construction contracting firms,[2] but the concepts presented are equally applicable to heavy construction. It is written for contractors that either do not use estimating software or use packaged software that frequently does not include all the items of work that they are called on to do. It assumes there is some pencil pushing during development of the estimate. It also assumes you know how to do a takeoff and estimate the labor and materials necessary to

---

Throughout this book I assume that the estimator is preparing a fixed-price bid for the work under consideration and that this bid is to be submitted to an "owner," who may be a person or organization, public or private. This assumption does not interfere with other proposal models such as negotiated contracts.

1. Marla McIntyre, "Why Do Contractors Fail?," *Construction Business Owner*, 62–65.
2. The term "small contractor" will be used throughout this book to mean a contractor with a full-time workforce ranging from several workers to — on occasion, as workload increases — approximately thirty or so tradespeople, not counting subcontractors.

do the job at hand. Throughout this book, there is an assumption that the construction company's owner takes a hands-on role in most estimates. This book is not in competition with any software you might buy. Rather, it provides a process for you to develop your estimate from the time you receive a request for a bid until you submit your bid. It is a guide developed from over 40 years of experience as a civil engineer and is intended for use in at least three ways:

1. To serve as a system of checks and reminders for working estimators
2. To assist people who are learning the craft of estimating
3. To assist those who teach estimating in a formal classroom setting

Generally, an owner will prepare estimates at different stages of project development to establish the baseline of the project's cost. These estimates are referred to as "design estimates." They are frequently used to determine the financial feasibility of the project and to obtain financing for it. Design estimates run parallel with planning and design as follows:

- Order-of-magnitude estimates
- Preliminary or conceptual estimates
- Detailed estimates
- Architect or engineer's estimate based on plans and specifications

Your contractor's estimate, or "bid estimate," represents the most detailed and accurate estimate of construction costs. The owner assumes you are aware of the market as it exists at the time of submission and as it will be during the period of construction. The bid you submit to the owner either for competitive bidding or for negotiation must include your indirect and overhead costs plus the direct costs of construction, including supervision plus markup and profits. The direct cost of construction for bid estimating is usually derived from the following:

- Quantity takeoffs
- Subcontractor quotations
- Company experience

Once you are selected as the successful contractor, your bid becomes your budget and "control estimate." Your firm will use your estimate for control, planning, and obtaining construction financing. The budgeted costs should also be regularly updated to reflect your estimated cost to completion and monitor cash flows, and you will want to compare your estimate against actual costs in preparation of your next estimate.

The process of estimating the cost of a construction project is logically divided into five phases:

1. Initial assessment
2. Work analysis
3. Programming
4. Costing
5. Cost distribution and summarization

The chapters of this book follow this outline.

1. INITIAL ASSESSMENT

Before starting an estimate you need to get acquainted with the project and its environment and establish a background of basic information from which your estimate will be developed. Your initial assessment should consist of the following fact-finding assessments, as applicable, and any others that you may find necessary or advisable. You may be tempted to jump ahead and get started with the estimate right away. Don't do it. A careful initial assessment will save you considerable grief. Furthermore, much of this initial information will undoubtedly be needed when you submit your application for a building permit; in the long run, avoiding an initial assessment won't save any time. When making your initial assessment, consider the following:

- Study all plans and specifications. Make sure they are complete. As the estimator you will essentially deconstruct the building. You should do this with an eye that peers beyond simply developing a bid. You are also a plan checker. As you study the plans and specifications, consider constructability, building codes, and the workers who must read the plans.

- Always inspect the work site and its vicinity. Consider especially safety, security, and access for equipment and materials and their storage. Determine the difficulty of providing and maintaining access to the project and other ways that the project's location may affect cost. Discuss any encroachments on traffic with local police and fire departments. Learn what special permits and fees are required.

- Obtain Department of Building and Safety records (e.g., previously issued permits, certificates of occupancy, and outstanding violations). Learn about local permit, inspection, fence, sign, and security requirements and relevant code and zoning regulations. If the project warrants, contact other local agencies: Public Works, Transportation, Planning, Zoning, Housing, and so on. The cost of building permits varies with the locality in which the work is being done. Some localities base the cost of the permit on the total cost of the project or the cu. ft. of the building, while others charge for water consumed, street frontage occupied, and the number of plumbing fixtures.

- Obtain soil and geological reports and any other professionally developed reports on the building and property.

- Take photographs that may be of value as you develop your estimate in your office.

- Discuss the project with the owner or owner's representative. Make sure there is full understanding between you and the owner. This may require considerable time and patience. Keep in mind that you may be speaking with a person who can't read plans, is not knowledgeable about construction, and is emotionally involved in the project.

- Collect wage and benefit rates for all the skills that will be used on your project. Anticipate wage hikes. If you anticipate hiring and are a union shop, meet with local union representatives to establish a working relationship. If you are a nonunion firm, seek out local trade associations; companies that provide temporary labor (some are specifically geared toward construction); and federal, state, county, and city agencies created to help people find work. If this is a new area, make a special effort to learn about the local labor market, including wages and benefits, working rules, and customs and worker attitudes.

- Check on working rules. For example, in some areas you may be required to provide 1 foreman for every 10 workers and in another 1 foreman for every 4 workers. If you must import workers, the imported workers may be entitled to a travel allowance. Some contracts require that all workers get the same pay. In some areas this may mean city workers who come onto the jobsite; these people may earn more money than your employees.

- Most construction projects involve some overtime work in addition to that which is already allowed in the trade labor items. For example, you might have to finish a concrete pour after normal work hours resulting in unanticipated premium pay. Payroll insurance rates are not paid on premium time, so either carry the item on the material side of your estimate or delete it from the gross labor costs before computing insurance costs.

- Know what bonds will be required:
    - Performance bonds guarantee the faithful performance of all work required to complete the contract. Payment bonds guarantee the payment of all bills for labor and materials used in the work, including materials purchased for the project but not included in it.
    - Bid bonds guarantee that the contractor, upon being declared the successful bidder, will enter into a contract with the owner for the amount of the submitted bid and will provide contract bonds as required.
    - A maintenance bond guarantees the performance of the contract. It covers the time after the building is completed. This bond stipulates the building stability and maintenance responsibility of the construction company after the building is finished.
    - Permit bonds are required by some local authorities to indemnify the authority against any costs arising from the contractor using public streets, crossing curbs, connecting to public utility lines, and so forth.
    - Supply bonds guarantee the quality and quantity supplied to the owner. These are rarely used on construction projects.

- Contact potential vendors and subcontractors. Ask for information about their firms and alert potential vendors and subcontractors to any requirements you may have before you enter into a contract. Also, contact communication and utility companies to learn about their requirements during construction. Obtain fees and rates.

- Have a clear understanding about all owner-supplied materials and equipment, their scheduled delivery dates, and on-site storage.

- Talk with knowledgeable local residents and building occupants to discern unique aspects of the site.

- Consider the weather over the life of the project.

- Learn all the owner's requirements to qualify as a bidder and all the owner's requirements that must be observed in bidding. Ensure the owner's financial and legal competence as they pertain to the contract. Ascertain the owner and owner's representative's technical competence. Make certain of the time allowed to complete the project and any penalties for late completion.

- Determine how you will distribute mobilization and demobilization costs. You will distribute these costs during the cost distribution and summarization phase of your estimate. Nevertheless, before starting your estimate it is wise to decide on how you will distribute mobilization and demobilization costs especially if the project requires a substantial amount of equipment. There are three common approaches:

  - Treat mobilization as an indirect expense, and absorb it on the basis of total direct costs or total direct labor costs.

  - Charge mobilization through a suspense procedure that assigns costs to direct expense.

  - Separately aggregate from both direct and indirect costs, in which case mobilization must be added to the various category totals essentially as a bid item. Some owners object to paying separately for mobilization because they cannot visualize mobilization as work that adds value to their project. Nevertheless mobilization is a very real cost to the contractor. Therefore, you must then distribute mobilization costs to other work items, usually to the bid items that benefit from the mobilization.

Collecting this information may be time-consuming and expensive, especially if you are moving into unfamiliar territory, area- or project-wise. But by not collecting this necessary information and attempting to take a shortcut, you open yourself up to making critical errors in your estimate and thus your bid, which could literally destroy your company. Therefore it behooves you to carefully plan and execute this initial phase of estimating and do it right.

Over time you will create your own initial assessment outline and standardize it. This will make the collection of information easier and the information comparable with other projects. The following checklist is provided as a starter.

INITIAL ASSESSMENT CHECKLIST

- General:
  - Project name.
  - Official name.
  - Identification number.
- Location:
  - State, county, city, town, and so forth.
  - Elevation.
  - Special site conditions.
- Owner and owner's representative(s):
  - Name and address.
  - Financial responsibility.
  - Officers and representatives.
  - Key personnel (e.g., architect, structural engineer, landscape architect, building department plan checker, etc.). This list will grow as more people are contacted

(e.g., local officials, labor representatives, material suppliers, equipment rental providers).

- Description of project. Briefly describe the project as you would
  on your application for a building permit (e.g., "Convert existing two-story, one-unit apartment building into a single-family
  residence. Add 1500-sq.-ft. attached addition to rear of house.").

- Applicable codes and plans and specifications, including where,
  when, and how to obtain these.

- Bidding and proposal information:
  - Bid or proposal due date.
  - Where and when bid or proposal is to be delivered.
  - Whether bid or proposal opening is public or private.
  - Form of bid or proposal. Is this a fixed-price bid or is it
    subject to further negotiations?
  - Bonding requirements (e.g., bid bond, contract bond,
    performance bond, payment bond, certified check,
    etc.) and amount required.
  - Time limits and any penalties and bonuses.
  - Auxiliary information that may be requested.
  - Contractor's (your firm's) financial report.
  - Your firm's experience record.
  - Your firm's in-house staffing, equipment, and key
    personnel.
  - Your firm's references.
  - Other evidence of your firm's fitness for the work.

- Physical factors (e.g., topography, hydrology, and geology):
  - Potential construction problems:
    - Drainage and groundwater. Are storm sewers pro-
      tected from sedimentation? Are manholes, inlets,
      and catch basins set to avoid ponding? Are they free
      of debris?
    - Rock outcroppings and large rocks that may have to
      be removed.
    - Steep slopes. Many jurisdictions have specific build-
      ing code and zoning regulations that will have to be

       met before a permit will be issued for work in steep slope areas.

- Whether loaded trucks can enter and exit easily.
- Distance to utility hookups. Will utilities have to be relocated?
- Will adjacent structures be affected?

- Weather interference. If you are moving into a new area, do not assume that knowledge of the general area is good enough. In many parts of the country climates vary drastically (microclimates) across very short distances:

  - Excessive heat and cold.
  - Sudden changes in temperature.
  - Excessive precipitation or none at all.
  - Excessive or deficient humidity.
  - Mud.
  - Frozen ground.
  - Fog.
  - Snow. Include an estimate for snowplowing and the effects of melting snow.

- Other considerations:

  - Catastrophic events such as storms, earthquakes, fires, flooding, and landslide hazards.
  - Ground water or no water. Obtain data on water levels and determine if they will affect your project.
  - Porosity and probable inflow of water.
  - Stability and expansibility of saturated soils.
  - Swampy ground.
  - Natural gas.
  - Insects and animal life. Are they destructive, annoying, or disease bearing?

- Leadership in Energy and Environmental Design (LEED) certification. Many owners are demanding that their buildings be LEED certified. Owners seeking a LEED certification will usually plan a course of action that includes preserving as much of the site as possible and then landscaping with native and adapted plants. Therefore it is imperative that you coordinate closely with the owner and architect as you develop your estimate, as pricing may vary from standards of the area. Some site decisions that would normally fall to the contractor may already have been made; price is very often not the main consideration for LEED-certified buildings. What follows are issues that may have to be considered as you price. The list is intended to give you a sense of what is involved, but it is not comprehensive. Again, you must work closely with the owner and architect or have ironclad specifications when estimating a building seeking a LEED certification:[1]

- Pollution. Identify measures to prevent erosion and sedimentation and prevent air pollution from dust and particulate matter. Assume that you will have to restore eroded areas.

- Land features. Identify bodies of water, exposed rock, unvegetated ground, or other features that are part of the historic natural landscape within the site, and learn from the owner the extent to which they will be preserved.

- Erosion. Identify areas of erosion and document them so that blame for them cannot be placed on the contractor.

■ Direct, indirect, and overhead costs.[2] These costs will be discussed in greater detail in the Chapter 4. However, at this initial stage

---

1. To learn more about LEED and green construction, contact the US Green Building Council, 1800 Massachusetts Avenue NW, Suite 300, Washington, DC 20036, http://www.usgbc.org.

2. "Direct," when applied to construction, means costs that can be specifically identified with a project or with a unit of production within a project. "Indirect" cost is cost that can be identified with projects but not any specific project or unit of production. "Overhead" costs are costs that cannot be identified with or charged to projects or units of production except on some more or less arbitrary allocation basis. These are costs that are incurred in the home office. Examples include accounting fees, insurance, home office telephone service, advertising, bid preparation for jobs that were not won, and other office personnel.

it is good to be aware of how they fit into your estimate. Most small contractors *distribute* direct, indirect, and overhead costs to all projects, but labor, rental, intangibles, expendables, and subcontractor costs can be identified with the specific project:

- Labor performed on the project. This is the labor that you estimate during the costing phase of your estimate and will be based on your takeoff.

  □ Estimation. When estimating the cost of direct labor, include the pro rata share of payroll taxes and related employee benefits such as workers' compensation; group insurance; holiday and sick leave; unemployment insurance; Social Security; and general welfare benefits such as health insurance, vacations, severance pay, pensions, profit sharing, and so forth. These are costly. You don't want to learn of some unexpected required benefit payments after your bid has been submitted.

  □ Labor and productivity rates. Become intimately familiar with your firm's labor and productivity rates. Learn the present wage scale for regular hours; shift work; daily overtime; weekly overtime; and Saturday, Sunday, and holiday work and overtime. Predict changes during the life of your project. Understand your firm's policies regarding work breaks and the meaning of start and completion time (e.g., does it mean dressed and ready for work at the start of the day or does it mean on site with time to get ready?).

  □ Area work rules. Learn the rules of the area with regard to limitations on work, crew sizes, hiring and discharging, and reporting to work. Learn if there are special rules on how to settle worker grievances. Are there special rules pertaining to payroll procedures? Is the contractor obligated to provide parking? Parking can be an especially onerous requirement when working in an inner city. Most construction projects will involve some overtime in addition to that which you might estimate. For example, you might want to complete a concrete pour at the end of a day so as not to have a joint. Therefore, you may want to add an item for this unanticipated overtime. Payroll insurance rates are not paid on premium time, so you can deduct it from the gross labor cost before computing insurance costs.

- Equipment rental and associated equipment costs:
    - Availability of rentals. Identify local companies that rent tools and equipment. Obtain price lists. Obtain references and learn how well equipment is maintained.
    - Equipment costs. Assuming you keep good records, keeping equipment costs is fairly simple. Whether the equipment is owned or rented, all relevant costs are charged to an account called "cost of equipment operation" or something similar. As the equipment is used on the job, its use is charged to the cost account on an hourly basis. This is usually referred to as "rentals," and estimators usually use the term "rental" when accumulating these costs. Under normal operating circumstances your accountant will assign more money to the work items than to the "cost of equipment operation" account. This resulting balance will later be used to absorb the cost of overhauls and other extraordinary repairs.
    - Rental costs. There is much to consider when estimating the rental of equipment, including the following points. These points should be covered in your equipment rental contract. Do not assume that the rental company has your best interest at heart:
        - Rates and rental periods. Usually equipment is rented on a monthly basis. Rental for fractions of a month are usually stated in days, and fractions of a day are usually stated in shifts or hours. When the

contract is stated at a flat rate per month, it is usually up to the contractor to determine usage and to keep the equipment running. If special attachments are needed, it is important to know if there are extra charges for these attachments.

☐ Light versus heavy equipment. For light equipment that is rented for a relatively short time, the rental period customarily starts when the equipment leaves the rental company's yard and ends when it is returned. For heavy equipment that is rented for longer periods, the rental period customarily begins when the equipment is loaded for shipment and ends at a stated time after notice of termination.

☐ Special provisions. Some rental agreements have special provisions that cover your liability for rent during time lost through strikes, storms, or other reasons that would be out of your control. Some contracts terminate the contract entirely, whereas others provide for a standby rate (e.g., half the standard rate). You may wish to ensure that you have some protection for such events.

☐ Move-in and move-out. The cost of moving the equipment from the rental company's yard to your job and back again is usually borne by you, the contractor. However, a costing problem arises if equipment is being moved from another contractor's project (i.e., not from the rental company's yard) to your project, and the distance is farther than from the rental company's yard. Make sure that the distance for which you are paying does not exceed the distance from the rental company's yard to your project. Some rental companies try to collect twice for equipment moving from one contractor to another; therefore you should get your rental company to certify that it is not charging the previous contractor for returning the equipment and you for having it delivered.

☐ Loading and unloading. Usually you will have to pay for unloading and reloading at your project, but the rental company does not charge for unloading and reloading at its yard; make sure this is so. Also, verify these fees when moving equipment from another contractor's project to yours.

☐ Assembling and dismantling. If you are renting a piece of heavy equipment that requires assembly after it arrives on your jobsite and disassembly before it leaves, the costs can be steep. These costs will normally be your responsibility. Equipment that arrives at your project in such poor condition that it cannot be operated should not incur rental costs until necessary repairs are completed. The rental company should pay the cost of these repairs, but then you will be required to maintain the equipment.

☐ Insurance. Most equipment rental contracts require the contractor (i.e., the user) to insure the equipment against accidental loss or damage.

☐ Idle time. Whether owned or rented, it costs money whenever equipment sits idle. If your schedule requires that a piece of equipment sit idle for a period, you still need to estimate the out-of-pocket costs of ownership, including taxes, insurance, depreciation or rent, and sometimes storage.

■ Materials, supplies, and other items that are actually consumed on the job, including related sales and use taxes:

• Project-specific costs. As the estimator you will be looking at the project's cost from the company's, not the project manager's, perspective. Therefore, costs such as fuel, oil, drill steel and bits, welding rods, utilities, communication, permits and licenses, small tools, move-in and move-out, security guards, storage of materials, and so forth that can be specifically identified to the project should be lumped and added to your project's price. The project manager who expects to be assigned to the project if the bid is successful may argue that these costs should be considered company-wide overhead and distributed as such across all ongoing projects. In the long run it is best to stick with a policy that assigns costs that can be specifically identified to a project.

- Supplies. Identify sources, availability, suitability, delivery times and cost of all commonly used building materials (e.g., rebar; fuel and lubricants; lumber; ready-mix concrete or cement and aggregates; brick, blocks, and stone; mechanical and electrical items; and building accessories). When all is said and done, each project is unique. You will have to create your own list of needs and then search for them.

- LEED certification. "Green" and "sustainability" have become important buzzwords. Many owners want to get their buildings LEED certified, thereby making them more attractive as rentals and usually more efficient in the long run. When qualifying materials and equipment are installed in a building, owners are granted points toward their LEED certification. But before a LEED certification will be awarded, all claims about sustainability will have to be verified; both new and existing buildings (additional categories are continuously being added) may be LEED certified.[3]

---

3. Verification procedures may rely on product certifications such as Green Seal and ENERGY STAR. Take care to confirm the validity of any product certification criteria before including it in the sustainable purchasing policy. An acceptable way to achieve this prerequisite is by using the US Environmental Protection Agency's Environmentally Preferable Purchasing (EPP) Program guidelines. The EPP Program information can be found on the associated website: http://www.epa.gov/epp.

Therefore prepare to carefully document all purchases. This documentation will probably go beyond your normal record keeping and may add cost to office operations. Also, pricing these materials will probably take more time than your normal material pricing.

Here are some things to consider as you price materials for a LEED-certified building. As with the LEED-related site issues, the list is intended only to give you a sense of what is involved, but it is not comprehensive. As you develop your materials estimate, you should work closely with the architect and engineer, as some of the items they specify may have no alternatives. Look for sources for the following:

□   Materials, supplies, and equipment that contain postconsumer and postindustrial material.

□   Rapidly renewable materials.

□   Materials harvested and processed or extracted and processed within 500 miles of your project.

□   Materials that may be salvaged from on-site. Do not presume that the only way or the cheapest way of disposing of vegetation is putting it in a landfill. Where markets exist, plant material from trees, grasses, or crops may be converted to heat energy to produce electricity. Timber can be sold for lumber or perhaps for composite wood or as firewood.

□   Forestry Stewardship Council (FSC)–certified wood.

□   Adhesives and sealants that have a volatile organic compounds (VOC) content less than the current VOC content limits of South Coast Air Quality Management District (SCAQMD) Rule #1168 or sealants used as fillers that meet or exceed the requirements of the Bay Area Air Quality Management District Regulation 8, rule 51.

- ☐ Paints and coating that have VOC emissions not exceeding the VOC and chemical component limits of Green Seal's Standard GS-11 requirements.
- ☐ Noncarpet finished flooring that is FloorScore-certified and constitutes a minimum of 25% of the finished floor area.
- ☐ Carpets that meet the requirements of the Carpet and Rug Institute (CRI) Green Label Plus Carpet Testing Program.
- ☐ Carpet cushion that meets the requirements of the CRI Green Label Plus Testing Program. Identify sources of composite panels and agrifiber products that contain no added urea-formaldehyde resins.[4]
- ☐ Lighting that meets or exceeds the very low levels of mercury usually specified for LEED-certified buildings.
- ☐ Materials that can be deconstructed, salvaged, and reused, when building codes permit.

- Often overlooked direct cost items. Develop a list of items that might be overlooked but should be charged directly to your project. Examples of these are warehousing, freight services, electricity, scaffolding, water, telephone, portable toilets, sewerage, and cleanup. There should be a clear understanding about whether the owner or contractor is responsible for utility costs. Make sure sufficient capacities are available to support your project.

- Subcontractor costs. Often, small contractors select subcontractors carelessly by limiting their search to familiar firms. If you are doing this, you may not be getting the best bargain for your firm or your customer. Chapter 4 contains an approach that gives you the opportunity to carefully select the best subcontractor for each project. It is based on a decision-making model

---

4. Composite wood and agrifiber products are defined as particleboard, medium-density fiberboard (MDF), plywood, oriented-strand board (OSB), wheatboard, strawboard, panel substrates, and door cores.

that is presented in *Construction Crew Supervision: 50 Take Charge Leadership Techniques and Light Construction Glossary*.[5] Other considerations include the following:

- Laws. Understand the laws, ordinances, and customs as they pertain to your project, and understand the liability your firm has for injury and damages to your employees, the public, the owner, the owner's employees, third parties and their employees on the project, and the project itself.

- Insurance. Learn what insurances are required, permitted, and available. The cost of insurance is generally added to the final estimate as an indirect cost, not including premiums that can be identified easily with a project, such as workers' compensation insurance and other mandated insurances that are tied to payroll costs. These are added to your bid as direct costs. Builder's risk insurance is also usually placed on individual projects. See Chapter 6 for additional information about insurance.

- Accidents. It is not possible to predict the cost of accidents. Therefore individual projects are not normally charged with the portion of the cost of recent accidents that is paid out as workers' compensation insurance until these costs are borne by the company in the form of higher premiums in later years.[6] Some companies, however, do charge projects when they close them out with either actual costs of accidents or some standard amount for each type of accident. This effectively raises

---

5. Karl F. Schmid, *Construction Crew Supervision: 50 Take Charge Leadership Techniques and Light Construction Glossary* (New York: Momemtum Press, 2009), section 32.

6. John G. Everett and Peter B. Frank Jr. concluded that "the costs of accidents and injuries have risen from a level of 6.5% of construction costs in 1982 to in between 7.9% and 15% today." This is an alarming trend. (John G. Evereett and Peter B. Frank Jr., "Costs of Accidents and Injuries to the Construction Industry," *Journal of Construction Engineering and Management* 122, no. 2 [1996]: 158–64.)

the perceived cost of accidents as perceived by the project manager. This model predicts that project managers will spend additional money to reduce future accidents. This additional cost can be predicted and added to your bid. More importantly, by charging projects directly as they progress, small contractors can evaluate the true effectiveness of project managers. Getting jobs done on time and within budget is good unless project managers cut corners, with resultant accidents coming to haunt the firm when workers' compensation rates rise.

- Taxes. Learn what taxes are levied on labor, purchases, property, business operations, and income. Learn if a special business license is required. When taxes can be identified to a project (e.g., sales taxes on materials), they should be deemed a direct cost. If you are estimating a government job, determine what, if any, purchases will be tax exempt. Obtain tax-exempt authorizations for your project.

- Licenses. Most jurisdictions require a license to do business within their boundaries. You may also have to obtain approvals to work from any number of public agencies, each of which may have a fee associated with it. Don't let yourself be surprised; do your homework and learn what permits and licenses you will have to obtain and the fees associated with them. Make sure there is an understanding between your firm and the owner as to how these fees will be paid. These permits would normally be indirect costs unless they must be obtained for a specific project.

- Permits. Building permits are required by most local jurisdictions. Therefore it is generally wise to establish a relationship with the local building department (also called the department of building and safety, buildings department, etc.) even before you have a job to estimate. These permits would normally be direct costs.

- Agency approval. Usually, before the building department issues a permit, it will make certain that all other government agencies have approved the project. Obtaining these approvals is the responsibility of the owner or contractor, not the building department. The list can be long: public works, transportation, planning, zoning, housing, schools, and special zones (e.g., coastal commission, environmental protection, and historic preservation), to mention some but probably not all. Public hearings and lengthy review periods may be required to obtain some approvals. Learn what approvals are required and know those people who are responsible for getting them. The cost of obtaining these permits would normally be directly charged to the project.

- Expenses. Bidding and business development expenses can often be identified to specific projects, and therefore the project should bear these costs. However, there are usually a number of unsuccessful bids for every successful one, and it certainly would be unfair to burden the first successful bid with the string of unsuccessful bids. Similarly, it would be unfair to allocate promotional expense and advertising to a single job. For these reasons bidding and advertising, along with expense accounts for managers, should not be allocated to individual jobs. They should be considered indirect costs and divided among projects.

- Office expenses. Home office charges are generally considered indirect costs. However, as the project progresses, the office accountant, as an example, may charge the project for the time spent in preparing the progress billing. When this procedure is followed, the accountant's time (and other office support) is charged directly to the project.

By the time your initial assessment is completed, you should have a full understanding of your project. You should know what the owner expects when you submit your bid, who the key players are, and who your subcontractors and vendors will be (or at least have a short list of candidates) and have them working on their estimates. You should also know about all regulatory, taxing, and other matters; the labor market and the peculiarities of the local area; and what your plan will be to complete your estimate.

2. WORK ANALYSIS

You do your takeoff during this phase of the estimate; I assume that you know how to do these quantity calculations. If not, there are many estimating books that provide this information. This book is not intended to help you with the mathematics required to obtain quantities but to help you determine just what information you need and how to use this information. Once you have done a takeoff (i.e., once you have determined the amount of work that needs to be done), there are three fundamental methods you can use to estimate the time it will take to do a task:

1. You can produce technical estimates using information that you purchased and your judgment based on your knowledge and experience. For example, from previous projects you know that it takes three carpenters and one helper 1 day to build and set up forms for a concrete wall 8 ft. high and 15 ft. long. You should certainly keep records of any work that you do repeatedly.

2. You can produce estimates that use statistical or historical data and time studies that you may have conducted. The data are based on the time it takes for a qualified person or group to do a task as compared to past records of similar people doing similar work under similar conditions. The data usually come from unadjusted work sampling or average historical production ratios. For example, if your work regularly entails concrete forms, you would want to develop statistical data on how many man-hours per sq. ft. it takes to construct rough forms for surfaces that are permanently hidden, ordinary forms for relatively unimportant exposed surfaces, good forms for normally neat exposed surfaces, and architectural

forms for highly important structures. You would also want to develop data on reusing forms.

3. You can produce estimates based on engineered performance standards that were developed from method-time measurements. These estimates involve tabulations of normal times for the increments that take repeat tasks together with allowances for local productivity, travel, material handling, job preparation, craft allowance, and personal allowance. Engineered performance standards (EPS) are used in estimating maintenance work for Department of Defense real property.[1]

Most small construction contractors use a combination of methods 1 and 2. They combine historical data with personal knowledge and experience to develop their estimates. In other words, estimates are compared with work already done and then priced accordingly. In one sense no two jobs are alike; however, most of your projects are made up of elementary construction operations that are common from project to project. The cost of the basic units varies only in degree based on the special aspects of each project.

Therefore, following your initial assessment, the next step is to break the work down into common terms. This is referred to as work analysis. Work analysis is more than a quantity survey (takeoff), although it forms a large part of it. It is a careful evaluation of every significant cost determinant that is part of the project. Most of the work analysis is done from a careful reading of the plans, specifications, and information you gathered

---

1. Engineered Performance Standards (EPS) are used in estimating maintenance work for Department of Defense real property: ARMY TB-420-4 through TB-420-30, NAVFAC P702.0 through P715.0, AIR FORCE AFM-85-42 through AFM-85-55. Manuals such as these can be purchased at http://store.ihs.com/specsstore/controller?event=LINK _DOCDETAILS&getCurVer=false&docId=KLJXDAAAAAAAAAAA&mid=W097. Companies that perform handyman service calls may find these manuals useful.

during your initial assessment. As already stated, you should not attempt an estimate without visiting the jobsite, during which you will want to take careful notes.

The following is a checklist of work items that will assist you as you do a detailed study of the plans and specifications. I assume that you already know how to do a takeoff and determine quantities. If you are using computerized estimating software (or your own records), you will be guided accordingly. For example, if your job entails the installation of 9 in. × 9 in. × 1/8 in. resilient tile flooring and the program you are using (or your records) gives an installation rate of 1.6 man-hours per 100 sq. ft., your takeoff would be in sq. ft. You may also choose to develop your takeoff based on how you must purchase materials or present your bid, or you may have to use both approaches. However you choose to do your takeoff, it is generally best to follow the same order, estimate to estimate. Carefully document your estimate, especially your planned sequencing, unit prices, and productivity rates, so the on-site project manager can learn how you arrived at your cost. The project manager's understanding of your estimate may help him arrive at a better way to do the work.

The checklist that is presented is lengthy, but it is still not all inclusive. Not every project will have every work item, but you may be surprised to find that more items than you expected need to be considered. Therefore I recommend reviewing the list prior to starting each estimate. There may be items that are not listed but that you identify as having to be estimated. Add them to your list. Your takeoff will give quantities (e.g., cu. ft./yd., board ft., tons, etc.) to which you will have to assign labor. Eventually costs will have to be assigned to all items, so organize your work carefully.

WORK ANALYSIS CHECKLIST

- Mobilization:
    - Plant and site setup:
        - Moving to and from site, loading, and unloading. Trace the route that you and your subcontractors will take getting to the project.
        - Clearing.
        - Excavation and grading.
        - Utilities. Will utilities have to be relocated?
        - Tool storage and work area setup.
        - Security.
        - Toilet facilities.
        - Heat.
        - Safeguards during construction.
    - Equipment installation:
        - Moving to and from the site, loading, and unloading.
        - Constructing equipment bases and supports.
        - Assembling equipment.
        - Setting equipment in place.
        - Connecting and testing equipment.
- Groundwork. Often your project begins here. Before beginning, make sure you have carefully investigated the site. Many contracts have exculpatory language that places responsibility for such an investigation squarely on the contractor. You do not want to learn that your site was once a landfill or that there is a rock outcropping just below the surface after signing the contract. Also be attentive to site access and preparatory work that may be required to keep equipment from getting stuck due to repeated traffic and rain.
    - Clearing. Make sure specifications are clear regarding matters such as the depth below finished grade that stumps and matted roots are to be removed.

- ☐ Trees. Make sure to mark and protect all trees that are to remain prior to starting your clearing operation.
  - ○ Area covered by trees.
  - ○ Type, size, and number of trees. Consider the type of roots of the various trees.
- ☐ Brush:
  - ○ Area covered by brush.
  - ○ Type, size, and density of the brush. Are there poisonous plants (e.g., poison ivy)?
- ☐ Grubbing.
- ☐ Area covered by stumps. Often this is the most expensive part of clearing and grubbing is the disposal of tree stumps.
- ☐ Type, size, and number of stumps.
- ☐ Root characteristics.
- Grass and weeds:
  - ☐ Area covered by grass and weeds.
  - ☐ Type and density of the grass and weeds.
- Disposal of vegetation:
  - ☐ Quantity of wood.
  - ☐ Quantity of brush, limbs, stumps, and roots.
  - ☐ Quantity of grass and weeds.
  - ☐ Proposed method of disposal. Do not presume that the only way or the cheapest way of disposing of vegetation is putting it in a landfill. You may have a saleable resource. Consider how the work will be accomplished.
  - ☐ Length of haul to disposal site.
  - ☐ Work associated with disposal.
- Removal of stones and boulders:
  - ☐ Area covered by stones and boulders.
  - ☐ Type, size, and number.
  - ☐ Degree of embedment.
  - ☐ Proposed method of disposal.

- □ Length of haul to point of disposal.
- □ Work associated with disposal.
- • Landscaping:
    - □ Area to be landscaped.
    - □ Type, source, and quantity of topsoil required.
    - □ Type, source, and quantity of sod required.
    - □ Area, type, and quantity of fertilizer.
    - □ Cultivation (plowing, harrowing, disking, weeding, etc.). Area and number of passes.
    - □ Seeding. Area to be seeded, type, and quantity of seed.
    - □ Trees and shrubs. Number, type, and size.
    - □ Care and protection. For example, some contracts require long term maintenance, sometimes up to 1 year of routine watering of trees. Do not downplay these costs.
- ■ Demolition:
    - • Quantities. It is helpful to do your demolition takeoff in the same units that you will use to develop your construction estimate. For example, if you are replacing gypsum board and you estimate that it takes 1.5 man-hours to remove 100 sq. ft. of gypsum wallboard, when you determine the number of sq. ft. of gypsum wallboard to be removed, you have also estimated the sq. ft. to be installed.
        - □ Plain concrete. Blasting, pneumatic tools, headache ball, drilling and rock jack, and so forth.
        - □ Reinforced concrete.
        - □ Asphaltic concrete.
        - □ Masonry.
        - □ Ornamental stone.
        - □ Tile.
        - □ Structural steel. Blasting, headache ball and torch, pneumatic tools and torch.
        - □ Metalwork. Finish and trim.
        - □ Partitions. Concrete block, brick masonry, cast-in-place, metal or wood studs, and so forth.

- ☐ Doors and windows.
- ☐ Flooring.
- ☐ Wallboard.
- ☐ Roofing. Corrugated metal, built-up, shingles, and so forth.
- ☐ Paving. Bituminous paving and base course, concrete paving and base course, curbing, catch basins, and manholes. Do signs need to be removed? Will traffic have to be directed?
- ☐ Piping and plumbing fixtures.
- ☐ Conduits, wiring, and electrical fixtures.
- ☐ Machinery.
- ☐ Piling.
- ☐ Overhead wires.
- ☐ Underground pipelines.
- ☐ Underground wiring.
- ☐ Fencing and guardrails.
- ☐ Rigging.
- ☐ Salvage. Gone are the days when demolished materials were simply sent to the dump. Identify materials that can be reused and recycled.
- Consider salvaging all building elements that are permanently and semipermanently attached to the building you are demolishing, including components such as wall studs, insulation, doors, windows, panels, drywall, carpeting, plumbing, electrical wiring, and flooring. Check with your local building department to learn what restrictions are placed on the reuse of these materials, especially lumber.
- Disposal of waste. Nowadays dumping may be the most expensive part of your demolition. Identify your disposal site with certainty. Do not underestimate tipping fees.
  - ☐ Determine the proposed methods of disposal.
  - ☐ Determine the length of haul to the disposal site.

- ▢ Determine the work involved in disposal. How will trucks be loaded and unloaded?
- ▢ Determine dumping fees and dump hours of operation.
- Earthwork:
  - Dewatering. Construction dewatering and water control are common terms used to describe removing and draining groundwater or surface water from a construction site by draining, pumping, or evaporating the water. Dewatering lowers the water table and generally is done before subsurface excavation for foundations, shoring, or cellar space. As this work is quite specialized, small contractors generally use a subcontractor for it.
  - Site drainage. Carefully consider scheduling this work. Scheduling sewerage and drainage lines early in your project may seem logical, but it will divide your work site, give you open trenches when you want to bring in materials, and leave you with loose pipes around the work site and generally restricted access. As you plan your work consider ways to prevent sediments from flowing off site.
    - ▢ Curtain drains.
    - ▢ Basement drains.
    - ▢ Sump pumps (pedestal or submersible).
    - ▢ Berms, retention basins, and so forth.
    - ▢ Some localities require a construction site storm-water management plan.
  - Excavation:
    - ▢ Grading and dozing.
    - ▢ Bulk earth moving.
    - ▢ Topsoil stripping.
    - ▢ Dozer-dug foundations.
    - ▢ Backhoe-dug foundations.
    - ▢ Hand digging.

- Trenching. Consider the methods you will use to trench (e.g., backhoe, hand dig, etc.).
- Fill and haulage:
  - □ Backhoe work.
  - □ Truck-in fill work.
  - □ Dozer work.
  - □ Handwork.
  - □ Tamping (area and layer depth).
- Septic systems. Consider both the septic tank and leech field. Make sure all permitting requirements have been met.
- Other earthwork, including embankments and backfill:
  - □ Processing required and waste material.
  - □ Wetting.
  - □ Spreading (depth of layers) and compaction (number of passes).
  - □ Hand placing and hand compacting of fill.
  - □ Material to be placed and compacted.
  - □ Conditions of placing.
  - □ Riprap and revetments.
  - □ Fine grading, slope trimming, and foundation preparation. Consider the area and type of work to be done.
  - □ Other hand digging.
- Materials:
  - □ Maximum and minimum sizes.
  - □ Quantity for each subsize. Check with quarry to ensure availability of specified sizes.
  - □ Take into account source, required processing, shrinkage and swelling, and waste and how it is measured for payment (e.g., bank measure, which is the volume of a given portion of soil or rock as measured in its original position before excavation; weight; or other).

- ☐ Consider how materials will be placed: by machine or by hand.
- ☐ Grout and mortar, if required.
- Drilling and shooting. Consider rate of drilling and cost of bits and steel. If explosives are involved, be sure to cover all safety and permitting requirements.
- Ground support. Estimate pressure to be applied versus soil carrying capacity, conditions of support, and design of support. Many local building departments have standard plans for this work, while others require a licensed engineer to design supports. Make sure you are knowledgeable of local practices.
- Underpinning. Underpinning is specialized work that involves danger to workers and structures. A professional engineer is generally required for design and often required to be on site during work. There are contractors that specialize in this work.
  - ☐ Magnitude of loads.
  - ☐ Condition of support.
  - ☐ Design of support.
  - ☐ Permitting and insurance requirements.
- Foundation testing. This is usually done by licensed professional engineers; however, you may be required to provide support labor to the engineer in the form of digging test pits, drilling, and so forth.
- Drilling and grouting. Drilling and grouting estimates can only be approximated. Quantities are estimated for bidding purposes, but substantial variations are common. Because of this, contract specifications and bid items are generally on a per unit basis. However, before subcontracting for this work, you should make a concerted effort to estimate the following:
  - Number, diameter, and length of grout holes.
  - Type of holes.
  - Characteristics of the ground, especially porosity.
  - Casing requirements.

- • Other considerations such as mobilization, environmental protection, and exploratory holes.
- ■ Anchor bolts or dowels set in holes. It may be less costly to drill anchor bolts than to lay out templates and install them during concrete pours.
  - • Holes:
    - □ Number, diameter, and length of holes.
    - □ Type of hole required.
    - □ Characteristics of material to be drilled.
  - • Bolts and dowels:
    - □ Number, diameter, and length required.
    - □ Required shop work.
    - □ Incidental materials (e.g., nuts, washers, etc.).
  - • Grouting materials. There are cement-based materials, resin-based materials, and epoxies available for grouting. Ideally your contract will specify the type of grout desired for the project. Some local building codes have grouting standards that must be followed.
- ■ Foundation drains:
  - • Drain holes:
    - □ Number, diameter, and length.
    - □ Type of hole required.
    - □ Characteristics of ground.
    - □ Casing requirements.
  - • Drain pipe:
    - □ Type, size, and length of sections.
    - □ Work required installing pipes.
    - □ Incidental materials (e.g., cement, aggregate, etc.).
  - • Porous fill:
    - □ Quantity in place.
    - □ Source.
    - □ Processing required.
    - □ Waste.
    - □ Shrinkage or swell. There will be a difference between the number of cu. yd. that are in place in,

say, a bank and the number of cu. yd. that you will have to haul. For example, dry loam may expand by 25% by the time you load it onto your truck. Many estimating guides provide swell factors for common materials.

- □ Quantity. Bank or quarry measure? Make sure your contract states how you will be paid (i.e., in situ measure or swelled measure).

- Paving. When estimating paving you should first speak with the owner about ways to reduce the heat gradient between developed and undeveloped areas. This can be done by using open-grid pavement on parking lots and walkways, choosing light-covered surfaces, adding reflective surfaces, and simply paving less area whenever possible. There are numerous software packages available to estimate paving, but beware of oversimplification.

  - Preparation of the subbase:
    - □ Base courses.
    - □ Number and thickness of courses.
    - □ Materials for each course.
    - □ Source.
    - □ Required processing.
    - □ Waste.
    - □ Shrinkage and swell.
  - Work required on each course:
    - □ Spread subbase materials.
    - □ Apply binding agent.
    - □ Compaction (and number of passes).
  - Bituminous paving:
    - □ Number and thickness of courses.
    - □ Aggregate for each course. Consider shrinkage and swell when purchasing aggregate.
    - □ Source:
      - ○ Processing required.
      - ○ Waste.
      - ○ Shrinkage or swell.

- ○ Method of payment
- □ Bituminous material. Type and grade:
  - ○ Priming and sealing coats for all courses.
  - ○ Work required on each course.
  - ○ Mix (in plant or on road).
  - ○ Spread (by hand, blade, or laydown machine).
  - ○ Compaction (and number of passes).
- □ Concrete paving (roads, driveways, and sidewalks):
  - ○ Thickness.
  - ○ Concrete materials.
  - ○ Forms. Type, size, and length.
  - ○ Screed, float, trowel, or machine finish.
  - ○ Waste.
  - ○ Method of payment.
- □ Concrete or granite curbs:
  - ○ Cross section and length.
  - ○ Forms. Size and length.
  - ○ Finish. Type.
- Fencing. Many projects require temporary fencing. Do not assume this is an insignificant part of mobilization to be covered by the project. Also, many cities have very specific regulations that specify temporary fencing construction.
  - • Type, height, and length of fence.
  - • Posts and tubing:
    - □ End posts (a.k.a. terminal posts). Number, type, and size.
    - □ Line posts (a.k.a. stretch posts). Number, type, and size.
    - □ Top-rail tubing. Number, type, and size.
    - □ Rail ends.
    - □ Tension bands.
    - □ Tension bars.
  - • Post holes:
    - □ Size and depth.
    - □ Characteristics of soil.

- ❑ Method of digging (e.g., hand or machine).
- Embedment of posts (e.g., earth, concrete, or other).
- Fence wire:
  - ❑ Mesh. Type, weight, width, length, number of rolls.
  - ❑ Individual wires. Number, type, size, length, and number of reels.
- Accessory materials:
  - ❑ Connectors, staples, and other fasteners.
  - ❑ Yokes, braces, and other post fittings.
  - ❑ Guys and anchors.
  - ❑ Gates.
  - ❑ Signs and signals.
- Trenching and pipe laying:
  - Trenching:
    - ❑ Characteristics of ground.
    - ❑ Methods of excavation (e.g., drilled and shot, backhoe, trenching machine, etc.).
    - ❑ Support.
    - ❑ Dewatering. Groundwater and inflow.
    - ❑ Backfilling and bedding. Source of material and work involved.
  - Trench support. The US Occupational and Safety Administration (OSHA) requires that trenches 5 ft. (1.5 m) deep or greater have a protective system unless the excavation is made entirely in stable rock. Trenches 20 ft. (6.1 m) deep or greater require that the protective system be designed by a registered professional engineer or be based on tabulated data prepared and/or approved by a registered professional engineer. There are different types of protective systems. Sloping involves cutting back the trench wall at an angle inclined away from the excavation. Shoring requires installing hydraulic or other types of supports to prevent soil movement and cave-ins. Shielding protects workers by using trench boxes or other types of

supports to prevent soil cave-ins. Many factors must be considered when designing shoring: soil classification, depth of cut, water content of soil, changes due to weather or climate, surcharge loads (e.g., spoil, other materials to be used in the trench), and other operations in the vicinity. Refer to OSHA and local building codes whenever trenching deeper than 5 ft., and in some localities 4 ft., in depth.

- Laying pipe. Materials:
  - □ Type of pipe (e.g., steel, cast iron, concrete, special, etc.).
  - □ Couplings, fittings, fixtures.
  - □ Material for joints.
  - □ Coatings and painting.
  - □ Work of installation. Weights of heaviest pieces, total weight, and special joints.
- Laying pipe. Installation:
  - □ Total weight to be handled.
  - □ Weights of heaviest pieces.
  - □ Type and size of couplings or joints, fittings, bolts, and so forth.
  - □ Type and length of welds.
  - □ Coatings and painting (both preparation and final).
  - □ Bedding and supports.
  - □ Dewatering. Groundwater and inflow.
  - □ Work involved in placing (e.g., hand tamping).
- Testing. Where and how taken. There are numerous ways to test piping: ultrasonic, hydrostatic, impact, metallographic, and so forth. Be sure to price testing correctly. Assume there will be some rejections and therefore corrective work.

■ Pole erection:
  - Holes:
    - □ Number, diameter, and depth.
    - □ Characteristics of soil.

- □   Proposed method of digging.
- •   Embedment:
    - □   Earth or concrete.
    - □   Materials required, if any.
- •   Guys:
    - □   Number, size, and length.
    - □   Anchors and fittings.
- •   Cross arms, insulators, and other pole fittings.
- ■ Well drilling. Usually, licenses issued to general contractors do not authorize well drilling; well drilling requires a separate license. So as a general contractor you will probably subcontract well drilling.
    - •   Subcontract considerations:
        - □   Mobilization.
        - □   Cost per ft. of drilling.
        - □   Cost per ft. of casing.
        - □   Cost of sealing materials and labor involved.
        - □   Cost of other materials (e.g., drive shoe, screen, perforated casing).
        - □   Cost of development.
        - □   Cost of pump test.
        - □   Cost of grouting.
        - □   Cost of pump and accessories.
        - □   Cost of disinfection.
        - □   Cost of testing.
    - •   If you are going to estimate the cost of drilling a well, the following need to be considered:
        - □   The records of existing nearby wells.
        - □   Diameter and depth of well.
        - □   Characteristics of ground and general formation.
        - □   Casing requirements.
        - □   Characteristics of aquifer.
        - □   Depth of first water.
        - □   Depth at which desirable water may be found.
        - □   Sections to be sealed and method of sealing.

- □ Sections of casing to be perforated.
- □ Treatment of aquifer.
- □ Required testing.
- Structural work:
  - Rigging:
    - □ Ropes and cables.
    - □ Sockets, shackles, couplings, and other fittings.
    - □ Sheaves, blocks, rollers, and other mechanisms.
    - □ Support structures and anchors.
    - □ Associated mechanical items (e.g., carriages, cages, bucket).
  - Falsework (i.e., temporary construction to support mortar, concrete, steel, etc.). You will probably be estimating based on a tentative design of the falsework. When you do your takeoff, consider reusing materials if allowed by code and union contracts.
    - □ Scaffolding.
    - □ Hoistways.
    - □ Walks and stairways.
    - □ Shoring and centering.
  - Concreting:
    - □ Check plans and specifications. Go beyond structural plans as concrete pads and so on are sometimes included in the mechanical and electrical plans. Is payment to neat lines or overbreaks?
    - □ Consider waste. When estimating for building construction, you usually do not have to deduct for small cuts, unless there are many. Have on-site project managers keep track of waste. For openers, estimate waste at 3% of total concrete volume.
    - □ Aggregates.
    - □ Classifications.
    - □ Materials of each class. Source and quantity.
    - □ Processing and required blending.

- □ Shrinkage and swell. Shrinkage is affected by curing conditions. (Concrete placed in 100% humidity for any length of time won't shrink.) Cure your concrete pour as long as possible.
- Reinforcing steel. There are many items to include in your materials takeoff: mesh, bars, stirrups, column ties, hoops, spirals, circle bent bars, bars bent, ties, supports, and so forth. If overlap has not been specified, you should generally add 10% to your estimate.
- Concrete forms. Designing concrete forms may be as complex as designing the structure you are building. When designing formwork do not exceed your knowledge.
  - □ For estimating, consider placing forms into categories:
    - ○ Function (e.g., footings, foundations, columns, walls, floor slabs, beams, girders, soffits, etc.).
    - ○ Surface type (e.g., rough or hidden surface, ordinary or good, exposed, architectural, etc.).
    - ○ Complexity (e.g., flat or large plane areas, straight with frequent parallel breaks, broken with frequent breaks in two directions, curved with straight lines in one direction, and warped with no straight lines).
    - ○ Single and multiple use forms.
    - ○ Form linings.
    - ○ Absorptive (used mostly on highways).
    - ○ Special board.
    - ○ Metal.
    - ○ Sculptured.
    - ○ Other.
- Concrete incidentals:
  - □ Expansion joint filler.
  - □ Mastic.
  - □ Caulking material.

- □ Bond preventatives.
- □ Water stops.
- □ Hardeners and accelerators.
- □ Air entrainment.
- □ Other as may be specified.
- Concrete finishing:
  - □ Formed surfaces to be pointed, patched, rubbed, or otherwise receive special finishes.
  - □ Unformed surfaces to be screeded, floated, broomed, trowled, or machine finished.
  - □ Construction joints and other surfaces that must be sandblasted, jetted with air and water, or otherwise prepared to provide a bond between courses.
  - □ Concrete curing.
  - □ Outside surfaces plus temporary joints.
  - □ Method of curing and curing materials (e.g., water, membrane, etc.).
- Embedded metalwork.
  - □ All items to be set in concrete and masonry.
  - □ Permissible tolerances in final position.
- Cold weather concreting. Pouring concrete in cold weather requires special attention and knowledge. You will need to heat the water and probably use accelerators. Concrete should never be placed on frozen ground; therefore the ground, as well as the forms, will have to be heated. You may have to keep the pour warm for several days.
  - □ Concrete materials.
  - □ Weight, aggregate sizes, and specific heat and temperature of each.
  - □ Temperature to which each must be heated.
  - □ Additives.
  - □ Heating the mix water. Total amount of water needed and raw water and heated water temperatures.

    ☐ Concrete protection.

    ☐ Include temporary joints in area calculations.

    ☐ Estimated heat loss and heating requirements.

    ☐ Type of protection (e.g., mats).

- Masonry:
  - ☐ Principal materials (e.g., bricks, blocks, stone, or tile).
  - ☐ Accessory materials (e.g., cement, lime, sand, color for mortar, antigraffiti materials, flashing, reinforcing, anchors, wall ties, inserts, bearing plates, lintels, support angles and channels, allowance joint pockets, waterproofing, etc.).
  - ☐ Complexity of work and pattern (e.g., straight and relatively free from breaks, badly cut up by doors, windows, etc.).
  - ☐ Cleanup (e.g., area cleanup, final cleaning, steam cleaning, acid cleaning, and power wash and pointing).
- Shotcrete and gunite. Shotcrete and gunite are commonly used terms for materials applied via pressure hoses. Shotcrete is mortar or (usually) concrete conveyed through a hose and pneumatically projected at high velocity onto a surface. Shotcrete undergoes placement and compaction at the same time due to the force with which it is projected from the nozzle. It can be impacted onto any type or shape of surface, including vertical or overhead areas.
  - ☐ Thickness to be shot.
  - ☐ Number of courses.
  - ☐ Reinforcing materials.
- Structural steel:
  - Field connections (e.g., bolts and welds).
  - Required working or annealing.
  - Testing.
  - Welder qualification tests.
  - Weld tests. Type and number.
  - Painting:

- □ Preparation of surface.
- □ Number and type of coats.
- □ Touch-up. Do not forget to include items that are painted prior to installation.

- Timber framing:
  - Treated and untreated timbers. Assume treated lumber wherever lumber comes in contact with concrete, masonry, and earth.
  - Bolts and connectors.
  - Nails and spikes.
  - Plates and other metalwork.
  - Framing. Cuts and holes.
  - Required treatment after cut and holes.

- Buildings:
  - Rough carpentry. Floors, walls including insulation and soundproofing, ceilings, and roof.
  - Millwork and finish carpentry:
    - □ Cabinets and built-in cabinets, countertops, shelving. Add cost to protect millwork and cabinets that are made by others.
    - □ Interior trim, panels, and ornamental work.
    - □ Stairs and components (e.g., balusters, newels, railings, etc.).
  - Doors, gates, and hardware. Metal, wood, special frames, and hardware. Be aware of the required fire rating for each door. It is usually wise to create a door schedule if the architect has not provided one. The schedule should include an opening number, door type, size, glass and louver requirements, and hardware. Pay special attention to unusually sized openings that require custom doors because costs cannot simply be prorated according to size. Identify all fire-rated doors.
  - Windows, unglazed. As with doors, it is usually wise to create a window schedule if one has not been provided. The schedule should include opening number,

window type, window size, glass type, frame materials and details, accessories and hardware. Nowadays some specifications require blast-proof windows for which no substitutes are allowed.

- Drywall and plastering. Wood or metal furring, gypsum board, DensGlass, insulation, corner bead, mud, tape, screws and nails. Do not forget scaffolding.
- Caulking, weather stripping, and water stops.
- Painting, staining, and varnishing.
- Building accessories:
  - □ Ventilators and louvers.
  - □ Gutters and downspouts.
  - □ Awnings, shades, blinds, and screens.
  - □ Mail chutes.
  - □ Vault doors.
  - □ Smokestacks.
  - □ Fireplaces.
  - □ Glazing (if not included with cabinets, windows, and doors). If tempered glass is required, price correctly.
  - □ Decorating (e.g., wallpaper, drapes, hangings, murals, statuary, signs, etc.). It is common for owners to provide these items. However, be clear as to who is responsible for installing them. Drawings should say words such as "Not in Contract," "NIC," or "By Others." If you are responsible, do not forget to include assembly and placement times in your estimate.
  - □ Floor coverings (e.g., carpeting, linoleum, tiles, terrazzo, etc.)
  - □ Specialty items. Your list of specialty items will vary from project to project. Many of these items are low cost but they can add up, so don't throw in a few dollars. Beware: Many specialty items are specified by manufacturer or catalog number and

substitutes may not be allowed. When receiving bids make clear whether costs include delivery and installation or not. Many specialty items require a support system (e.g., electric wiring and structural support). Be sure to account for these.

- Roofing and waterproofing (frequently done by roofing contractors):
  - □ Shingles, slate, tiles, metal, asbestos, paper, or combination of these materials.
  - □ Built-up roofing or membrane waterproofing.
  - □ Vapor barrier and insulation.
  - □ Skylights, roof hatches, and vents.
  - □ Other, as required by roof type (e.g., gutters, scuppers and downspouts, flashing, cant strips, expansion joints, roof pavers, gravel stops, fascias, reglets, termite shields, roof walkways to mechanical equipment, etc.). Some items may have to be prefabricated.
- Roofs that reduce heat islands:
  - □ High-reflectance materials to reduce the solar reflectance index (SRI) to meet Leadership in Energy and Environmental Design (LEED) standards.
  - □ Vegetated roofs.
- Elevators and dumbwaiters. Residences will usually have either a hydraulic or a machineroomless (MRL) elevator. Usually an elevator company will install the elevator, but you will probably be required to do ancillary work. For example, you will probably construct the shaftway, pit, and machine room. These will require lighting, ventilation, fire protection and smoke control, and electric service. If you are installing a hydraulic elevator, make sure responsibility for drilling is spelled out; the shaft must be drilled to a very close tolerance.

The same is true for elevator doors and floor doors. Throughout, work closely with the elevator company.

- Elevators are closely regulated. Make sure your plans have been reviewed and approved by the local regulating agency. There is usually a separate fee for this review that is on top of the building permit. Also, make sure all Americans with Disabilities Act (ADA) requirements have been met.

- Because elevators are not the usual fare for small contractors, inspection checklists have been added for both hydraulic and machineroomless elevators in Chapter 6, section 6.5.

- Mechanical, electrical, and plumbing (MEP) systems. Review all plans and specifications with an eye to these systems. Architects frequently subcontract for MEP and structural designs. This leads to breakdowns in coordination between drawings and specifications. Be on the lookout for such things as ducts conflicting with structural members, concrete pads for equipment not noted on structural drawings, and overall constructability (e.g., will the mechanical equipment fit into its allotted space). Document your findings.

- Fixture and device schedule. As with doors and windows, it is wise to create schedules for MEP, including controlling devices and accessories such as outlet boxes.

- MEP subcontractors will probably install the following systems (therefore scheduling is extremely important):
  - Plumbing. Coordinate all connections to local water and sewer lines with local agencies. If a tankless hot water heater is to be installed, make sure the gas service is large enough for the heater. Isolate the heater for noise deadening.
  - Heating, ventilating, and air conditioning (HVAC).

- □ HVAC controls. Coordinate the installation of all control systems between electrical and mechanical contractors.
- □ Electrical. Fix responsibility for structurally supporting electric panels and so forth. Determine what must be in place prior to the electrical subcontractor coming on the job.
- □ Vacuum system.
- □ Pneumatic system.
- □ Sprinkler system.
- Nonlisted items. Be alert to pricing and fixing responsibility for such work items as roof penetrations for kitchen and bathroom exhausts; dust protection, especially for renovations in occupied buildings; testing and balancing MEP systems; flushing pipes and ducts; and so on.
- Processing and operating equipment. Large machinery such as processing and operating equipment is usually viewed as a project unto itself.
  - Name, number, type, weight, and bulk.
  - Make, model, and catalog number.
  - Weight and bulk of subassemblies.
  - Incidental materials.
  - Required special services (e.g., assembly crew).
  - Testing and acceptance.

VALUE ENGINEERING

As you work through your estimate, you will be figuring out how to do the work and how to cut costs; you will probably not be thinking in terms of "value," but you should train yourself to do so. Value, as defined, is the ratio of function to cost. Therefore you can increase value by either improving the function or reducing the cost.

Value engineering (VE) is a systematic method to improve the value that you will provide to your client. It does this by an examination of function. Importantly, it is a primary tenet of value engineering that basic functions be preserved and not be reduced as a consequence of pursuing value improvements. Federal government and many state contracts require that you conduct a VE analysis of the project that you are bidding on. Savings are usually shared.

Value engineering is often taught within construction management, project management, and industrial engineering programs as a technique by which the value of a completed project is optimized by crafting a mix of performance (function) and costs. In most cases this practice identifies and removes unnecessary expenditures, thereby increasing the value for the manufacturer or their customers.

VE follows a structured thought process that is based exclusively on function (i.e., what something does, not what it is). It uses rational logic (a unique how-why questioning technique) and the analysis of function to identify relationships that increase value. For a more complete discussion of VE, see Chapter 6, section 6.6.

3. PROGRAMMING

Here the term "programming" is used to designate the process of deciding how and when to start each part of the work. This requires that you dissect your project into blocks of work (usually referred to as operations or tasks) of the size and form most convenient to you for the project you are considering.

Your project may be subdivided into a number of tasks necessary to complete the job. Each task may be performed by any number of possible combinations of methods, labor skills, crew sizes, equipment, and labor hours. Accomplishing each task at the lowest possible cost so that the overall project cost is also the lowest cost is usually the goal, but there may be other considerations. For example, an owner may be more interested in an early completion and utilization of the facility you are constructing. Of course, as a project drags on, your costs will also increase, and of course, customer dissatisfaction may cause you to lose future work. Therefore look at the big picture and consider all cost items connected to your project. When you do this you inevitably introduce time into your estimating equation, since almost all costs vary with time. But they tend to vary with time in opposing directions. Direct costs tend to decrease when you allow, within reason, more time to do the actual work, while indirect and overhead costs associated with the project's duration increase as the project lengthens. Most owners try to find this balance when they specify a completion date, and you will seek it as you plan your detailed schedule of operations.

The issue of balancing cost and time is not a new concept; however, accomplishing this balance is not easy. Each task into which you divide your project can be done in differently, resulting in different costs and lengths of time. Things would be easy if time did not matter, in which case you would perform each task at the lowest possible cost. But this is seldom the case. Generally you have to figure out ways to reduce project time from the lowest-direct-costs approach to a more favorable solution. Unfortunately, this usually adds to your direct costs (e.g., by increasing crew size and thereby working less efficiently, adding overtime and shift work, and adding to labor costs or bringing in larger, more expensive equipment).

The solution to this balancing problem would still be relatively easy if all tasks could be done in series—that is, if you worked on only one task at a time. Then you could shorten the project to its desired length by rushing the tasks that are the least costly to hasten. Alas, construction projects are almost always a complex network of concurrent and interrelated tasks. Therefore it is possible to rush a task and yet not affect project duration while still increasing project cost. Or you may choose to throw out the concept of balancing cost and time and rush all tasks. This would decrease project duration but increase project cost far more than necessary. In between are many possibilities (e.g., shortening a single task that may be expensive to shorten may be wiser than shortening a task that may be easy and low cost to change but affects other tasks). What is needed is a systematic method on which to base your decisions.

There are four main scheduling techniques used in the construction industry.[1]

---

1. Each technique is discussed in greater detail in Chapter 6.

3.1 GANTT (BAR) CHARTS
AND LINKED BAR CHARTS

A Gantt chart is a horizontal bar chart developed in 1917 by Henry L. Gantt, an American engineer and social scientist. Gantt charts are the easiest and most widely used form of scheduling in construction management. Even with other scheduling techniques available, the eventual schedule is usually presented in the form of a Gantt, or bar, chart. It provides a graphical illustration of the schedule. A typical bar chart is a list of activities with the start, duration, and finish of each activity shown as a bar plotted against a time scale. The level of detail of the activities depends on the intended use of the schedule. They may vary from simple versions done on graph paper to more complex automatic versions that are created using computerized project management applications. These can be found by searching "scheduling, bar charts" or a variation of this on Google.

The linked bar chart shows the links between an activity and its preceding activities that must be complete before this activity can start.

Bar charts are also useful for calculating the resources required for the project. To add the resources to each activity and total them vertically is called a resource aggregation. Bar charts and resource aggregation charts are useful for estimating the work content in terms of man-hours and machine hours.

Bar charts give a clear illustration of project status, but they do not show task dependencies (i.e., you cannot tell how one task falling behind schedule affects other tasks). The importance of this will become evident when other scheduling methods are discussed.

3.2   CRITICAL PATH METHOD (CPM)

The Critical Path Method (CPM) offers a systematic approach on which to base decisions. In my opinion it offers the most practical approach for a small contractor. Once learned, CPM is quite intuitive, and it provides an excellent visual display of how your project will proceed. CPM does the following:

- Pinpoints tasks whose completion times are responsible for your project's final completion time. This allows you to focus on these tasks and keep your project on time.
- Gives you as quantitative evaluation of just how much leeway each task, other than critical tasks, has. This leeway is called "float." Tasks with float may be started later or completed more slowly than your original schedule without jeopardizing overall project completion time. This knowledge is invaluable during project estimating and planning.
- Shows the most economical scheduling of all tasks for each possible project completion date. This is important because it permits you to correctly consider both time and cost when choosing methods, equipment, staffing and work hours, and materials. It replaces the trial-and-error juggling of task times to arrive at a possible solution.
- Provides necessary data for choosing the best completion date. The results of the most economical scheduling of all tasks can be plotted on a time-cost graph, along with curves for indirect costs and bonus-penalty or liquidated damage payments. The sum of these curves produces a total cost curve having a lowest cost point that indicates the optimum project duration.
- Determines the effect that a change in or addition of one operation will have on the project's duration. This is especially helpful when a client asks for change orders, as such changes often affect tasks that are not immediately apparent. This, in turn, helps as you negotiate price and project duration.

In Chapter 6, section 6.7, you can find the mechanics of a non-computer-based method for applying CPM. It is presented for the following three reasons. First, almost by definition, small contractors work on small projects, at least by construction industry standards. Once you learn how to develop a CPM schedule, you may find the non-computer-based approach to be more practical for a small project than using a computer; of course, this is a personal preference. Second, it offers you an opportunity to learn the details of the method and the assumptions on which it is founded. Third, after working through the details of the method, you will probably never approach work as you did in your pre-CPM days. In the spirit of full disclosure, after learning CPM you may drive your spouse crazy as you assist with housework and move from task to task in ways that are actually most efficient but may not be as she or he does them. Working through the nitty-gritty presented won't be easy, but you will be become a better, and certainly more competitive, contractor once you have mastered the concepts of CPM.

3.3   LINE OF BALANCE

Line of Balance is a planning technique used primarily for repetitive work on large projects, such as road and high-rise construction. In both there are tasks that are repeated across numerous work areas. Line of Balance is frequently used following the development of a CPM network.

Line of Balance shows the repetitive work of a project as a single line on a graph rather than a series of individual activities on a bar chart. It can be used for any project where there are a number of separate but common activities to undertake (e.g., foundations and roof construction that would be completed on each house of a multiple-unit development or an activity with a long duration). Line of Balance graphical methods result in a common understanding of how crews follow one another through the project. Such an immediate, intuitive understanding of the project is often not possible using other scheduling techniques.

Currently, some large firms are using Line of Balance scheduling, but I am not aware of any small contractor using it. Therefore what is presented is an introduction only. Research is currently ongoing, and you are sure to hear more about this scheduling technique in the future as more companies implement it and software is developed for it. Chapter 6, section 8, provides additional information about Line of Balance scheduling.

3.4 Q. SCHEDULING

Q. Scheduling is a new technique, though it is gaining popularity among contracting firms. Q. Scheduling has evolved in response to many contractors finding CPM too difficult to use, especially phase III. When required to provide a CPM network with their bids, many contractors hire a consultant to develop it, thereby losing the primary benefit of CPM: gaining a full understanding of the project.

Another driving force for Q. Scheduling is CPM's low weighting of cost control and efficient use of resources. Q. Scheduling reveals a relation between the sequence of doing a job and the costs to be incurred. The Q. Schedule is similar to the Line of Balance but is modified to allow for a varying volume of repetitive activities at different segments or locations of the construction project. An example is provided in Chapter 6, section 6.9.

At this point in your estimate you should have a pretty good understanding of what needs to be done and how long it will take. Your initial assessment gave you an understanding of the project and the issues confronting you. During the work analysis you completed your takeoff. This gave you the amount of work that will have to be accomplished and materials and equipment that must be rented or purchased. Programming gave you a schedule. We now get to the reason for being in business: money (unless, of course, you are wealthy and in the construction business for the fun of it).

4. COSTING

CONSTRUCTION ACCOUNTING OVERVIEW

Before delving into the topic of costing, a brief discussion on construction accounting is provided. This diversion has two purposes:

1. To help you understand where the numbers you use to develop your estimate come from and why you must add considerable dollars to each project beyond the direct cost of labor and materials implanted in your project. If nothing else, this may help you when you negotiate a contract and must answer to the comment "Don't tell me you pay your carpenters more than I earn as a bank vice president."
2. To help you discuss with your accountant just how to account for income and expenses.

Accounting for construction projects is particularly nuanced. Because of the long-term nature of construction contracts, the way you choose to report income and expenses associated with projects affects your financial statements and your taxes. Beyond this, understanding the numbers that make up your income statement, balance sheets, and cash flow statements may affect the way you bid each project.

A unique aspect of construction accounting focuses on just when profits are earned. This is not a problem for service calls and very small jobs that are finished quickly. In these cases the work is accomplished and the sale and revenue recorded. But many, perhaps most, construction contracts extend beyond one accounting period. Large contractors usually consider long-term contracts to be those that extend beyond 1 year, but as a small contractor you may consider a 3- or 4-month project to be long term. If you are a design-build firm—even a small one—the meaning of long term may be extended considerably. Whatever definition you and your accountant decide on, construction contract features such as target penalties and incentive awards, change orders, and so forth tend to complicate what for most industries is fairly straightforward accounting.

Long-term construction contracts usually require you, the contractor, to bill your client at intervals as various points in the project are reached. When the project to be constructed consists of separate units, such as two houses, provisions may be made for passage of title, as well as billing at identifiable stages of completion. The accounting profession recognizes two distinctly different methods of accounting for long-term construction contracts: the percentage-of-completion method and the completed-contract method.

1. Percentage-of-completion method. Revenues and gross profit are recognized each period based on the progress of construction—that is, the percentage of completion. Construction costs plus gross profit earned to date are accumulated in an inventory account (construction in process), and progress billings are accumulated in a contrainventory account (billings on construction in process).[1]

---

1. A contrainventory account reduces an asset, a liability, or an owner's equity on a balance sheet. Contrainventory accounts give the accountant some flexibility in presenting the financial information. In this case a reader of the statement could see construction costs plus gross profit earned to date in the construction in process account and progress billings in the billings on construction in process contrainventory account.

2. Completed-contract method. Revenues and gross profit are recognized only when the contract is completed. Construction costs are accumulated in an inventory account (construction in process) and progress billings are accumulated in an inventory account (billings on construction in process).

The percentage-of-completion method of accounting is based on the theory that profits are earned as the project progresses. It takes into account project revenues as a portion of the total contract price proportionate to the percentage of completion of the work. In contrast, the completed-contract method is based on the theory that no profit is earned until the project is completed.

The rationale for using the percentage-of-completion accounting method is that under most of these contracts the client and contractor have obtained enforceable rights. The client has the legal right to require specific performance on the contract; the contractor has the right to require progress payments that provide evidence of the client's ownership interest. As a result, the economics of the situation suggest a continuous sale occurs as the work progresses, and revenue should be recognized accordingly.

There are specific rules as to which method to use, and you need to discuss this decision with your accountant. However, usually the percentage-of-completion method should be used when estimates of progress toward completion, revenues, and costs are reasonably dependable, and the completed-contract method should be used when jobs are short term in nature, when the specific accounting rules prohibit use of the percentage-of-completion method, or when there are inherent hazards in the contract beyond normal, recurring business risks (i.e., when there is a chance either you or your client may not meet your contractual obligations for some reason).

From your accountant's perspective, making reasonably accurate estimates of completion and final gross profit for the percentage-of-completion method is extremely difficult. Your estimate and the units you use in preparing your estimate will play an important role in your accountant's estimates of project revenues as a portion of the total contract price proportionate to the stage of completion.

Some firms provide several different services or specialize in unique projects. In such cases these different services probably carry different profit margins based on their degree of risk, uniqueness, or simply your ability to charge more for certain kinds of work. For example, if you offer design-and-build services, the design portion of the project may carry a higher profit margin than the build portion. In this case your accountant may recommend segmenting the project as if the two parts are separate contracts despite the fact that you have sold both under one contract. You should discuss the benefits of segmenting projects and the accounting rules that pertain to segmenting with your accountant.

From this brief discussion about accounting, you should realize that your estimate goes beyond the project. It affects all aspects of your firm's life; indeed, one bad estimate can put a small contracting company into bankruptcy. There are really two ways this can happen: (1) by failing to include all quantities and usages in your estimate and (2) by assigning incorrect prices to the quantities estimated. In short, all aspects of the project must be assigned a cost.

COSTING: GENERAL

"Direct," when applied to construction, means costs that can be specifically identified with a project or with a unit of production within a project. "Indirect" cost can be identified with projects in general but not any specific project or unit of production. "Overhead" costs are costs that cannot be identified with or charged to projects or units of production except on some more or less arbitrary allocation basis. These are costs that are incurred in the home office. Examples include accounting fees, insurance, home office telephone service, advertising, bid preparation of jobs that were not won, and other office personnel. If you own the company, you will have to decide how to distribute these costs. As the estimator you will have to be given guidance on how much to add to each estimate.

However exact these definitions are, they may be applied differently depending on whether the costs are viewed from the home office or jobsite. From the project manager's perspective at the jobsite, all costs that cannot be specifically identified with a bid or work item are considered indirect, or "distributable," costs. These would include such things as fuel, compressed air, drill steel and bits, welding rods, and so forth, but when viewed from the home office, these would be considered direct costs because they can be identified with a project.

There are other costs that the home office would consider to be direct while the project manager would consider them indirect. For example, an accountant in the office may charge each project for the time she spends preparing progress billings. In addition, sometimes certain items of cost may be either a direct or an indirect cost. For example, a first set of aerial photographs made for a particular project might be a direct charge to a specific bid item such as site survey, but the next set of photographs might be for job progress pictures, which would be considered an indirect cost item. Resolve all these issues with your accountant. Ideally these issues should be resolved prior to starting your estimate.

These distinctions are not merely academic even for a small contractor because they may bear on contract pricing. This is especially true when the bid must identify individual bid items with unit prices. It becomes a major factor once the project is ongoing and actual costs must be tracked against the original bid for the purpose of projecting the final project outcome. It also becomes an important factor in pricing change orders and in the event of disputes between you and the owner over individual bid items within a project. These costs and price disputes often center on the allocation of indirect and overhead items among bid items. The term "costing" refers to the process of converting the work that you have done into costing items and assigning prices to them. The term "costing item" refers to specific items that are paid for by the project.

Costing items fall into the following categories:

1. Staffing
2. Equipment and equipment usage
3. Intangible expenses
4. Tangible items that are not incorporated into the work
5. Tangible items that are incorporated into the work
6. Subcontract cost

Costing items usually have a dual nature. They include the price per unit of quantity per unit of usage. For example, if a crew of six carpenters works eight 40-hour weeks, the quantity is "six carpenters" and the usage is "320 hours," but the unit price on which payment is based is "price per carpenter-hour." Exceptions to this dual nature concept would be, say, a pickup truck that is rented for the period of the project. In this case the costing item would be the single lump-sum cost of the rental. However, if you are listing costing items on a spreadsheet that includes columns titled "Quantity," "Usage," "Unit Price," and "Total Cost," it is usually convenient to place the number 1 under both Quantity and Usage and the lump sum price under Unit Price.

For small contractors usually the simplest and most satisfactory segregation of direct, indirect, and overhead costs is to distribute all components of cost except the following, which are direct costs:

1. Labor performed at the jobsite, including a pro rata share of payroll taxes and related employee benefits (e.g., workers' compensation, group insurance, holiday and sick pay, and other fringe benefits).
2. Materials and supplies that are actually consumed on the job, including sales and use taxes.
3. Rented equipment used on the job.
4. Subcontractor costs.
5. Department of building and safety and other agency fees that may be required for plan reviews, permits, and other clearances, including the time spent obtaining them. As a small contractor your contract may state that the owner will reimburse you for plan reviews, permits, clearances, and so forth, but this neglects the cost of time spent obtaining them. This cost of labor, usually of a key person, may be considerable.

Costing items are usually specified by the following:

1. Assigning a categorical name that identifies the item and to which a unit price is assigned (e.g., carpenters, $45.75 per hour).

2. Stating the number and name of the units of usage (e.g., six carpenters).
3. Stating the number and name of the units of usage to be applied to each unit of quantity (e.g., six carpenters for 320 hours).

Costing itself is done in two distinct ways:

1. By setting up a work unit of people and equipment (e.g., a dozer and operator), determining the productivity of the work unit, tabulating the costing items applicable to this work unit, and pricing them out on the basis of the time required to do the work.
2. By pricing the work directly on the basis of units of costing items required per unit of work.

Choose the costing method based on which is most applicable to the specific work in question and on the costing data available. Costing method 1 is usually used for complex operations, such as excavation and concrete pours, that have variable cost determinants. Costing method 2 is usually adapted to craftwork that maintains a relatively high degree of uniformity between projects.

When you are working on a spreadsheet, you probably will want to use one-word headings, which include the following:

1. *Labor.* All staffing costs, including payrolls, fringe benefits, workers' compensation insurance, Social Security, and all other expenses related to the use of staffing.
2. *Rental.* Depreciation, amortization, write-off, rental, or any other charge made for the use of owned equipment (but not operation), as well as rent paid to others for the use of equipment, including operation, if rented on an operated basis.
3. *Intangibles.* Freight, taxes that are not charged to the cost of goods or labor, insurance, bonds, services, and other expenses for which no material goods are received.
4. *Expendables.* Purchases of material goods that are used in the construction project but are not incorporated into the work itself.

Expendables remain the property of the contractor until they are used or disposed of.

5. *Material.* All material goods purchased by the contractor and incorporated in the project, thereby becoming the property of the owner.

6. *Subcontracts.* All payments to subcontractors.

Costing for small jobs can be done by hand, or you may create a spreadsheet. Your spreadsheet need not be complicated. For example, Figure 4.1 shows a simple form for recording costing items, the unit prices assigned, and costs. With some exceptions you must make two multiplications to obtain the cost. If your accountant wishes, you can set up your spreadsheet so that your calculations can be distributed under the one-word headings (i.e., Labor, Rental, and Materials). Figure 4.2 is an example of an easy to create costing spreadsheet that distributes costs in this manner. This spreadsheet also looks forward to when you will distribute and summarize your costs.

Other spreadsheet formats can be found in easy-to-purchase estimating guides.[2]

It is good practice to develop a routine by which you do your estimates. I have found that, for convenience and clarity, one should treat costing as four sections: mobilization, plant operations, direct expenses, and indirect expenses.

MOBILIZATION

Prior to starting a project, you need to move equipment and materials to the jobsite. You may also have to prepare the site by building sheds and fencing, moving in an office trailer and toilets, and so forth. All this preparatory work is referred to as "mobilization."

2. As an example, see Ed Sarviel, *Construction Estimating Reference Data* (Carlsbad, CA: Craftsman Book Company, 1993).

Figure 4.1. Costing sheet

HOME BUILDING AND REPAIR INC.
COSTING ITEMS

Project: 2345 Eastman Ave.   Page: 1
Estimated By: KFS   Date: XX/XX/XXXX

COSTING ITEM	QUANTITY NUMBER	QUANTITY UNIT	USAGE NUMBER	USAGE UNIT	COST UNIT PRICE[1]	COST TOTAL
Project manager	1	Each	12	Weeks	2,300.00	27,600
Carpenter, Foreman	1	Each	320	Hours	55.50	17,760
Carpenter, Journeyman	6	Each	320	Hours	45.75	87,840
Carpenter, Apprentice[2]	1	Each	320	Hours	35.00	11,200
Lumber	10	MBF[3]	1	–	203.50	2,035
Rented truck	1	Each	12	Weeks	205.00	2,460
Fuel for truck	8	GPW[4]	12	Weeks	3.05	293

Notes:

1. Wage rates include fringe benefits.
2. Ideally, the unit price for each item should remain constant. However, apprentice rates increase annually for 4 years, until the apprentice becomes a journeyman. Here, the average apprentice rate (dollars per hour) is used.
3. Thousand board ft. "Board ft." is a measurement of lumber volume. A board ft. is equal to 144 cu. in. of wood. It is easy to calculate board ft. using the following formula:

    (thickness × width × length) / 144 = board ft.

    Lumber is specified by its rough size. This is why a 1 in. × 4 in. board is actually ¾ in. thick and a 2 in. × 4 in. board is actually 1½ in. thick.

    When you are figuring up board ft., keep in mind a waste factor. If you purchase good, clear material, add about 15% for waste; if you elect to use lower-grade material, you will have to allow for defects and more wasted material, so add about 30%. Take a few boards and run the measurements and you will see how easily this works.
4. Gallons per week.

Figure 4.2. Cost distribution sheet

PROJECT

EST. BY   DATE: 2/28/2010   SHEET NO   OF SHEETS

| COSTING ITEMS | | | | | COST | | COST DISTRIBUTION | | | | | | | | | | | |
| NAME | QUANTITY | | USAGE | | COST | | LABOR | | RENTAL | | INTANGIBLES | | EXPENDIBLES | | MATERIAL | | SUBCONTRACT | |
	NUMBER	UNIT	NUMBER	UNIT	UNIT PRICE	TOTAL	U.P.	AMOUNT	U.P.	AMOUNT	U.P.	AMOUNT	U.P.	AMOUNT	U.P.	AMOUNT	U.P.	AMOUNT
Trench 1,500 LF																		
Machine	1	Mch	10	Day	1,200.00	12,000			1,200	12,000								
Operator	1	Wkr	80	Hr.	42.50	3,400	42.50	3,400										
Total Trench						15,400	42.50	3,400	1,200	12,000								

If trenching machine was rented with an operator it might look like this

| COSTING ITEMS | | | | | COST | | COST DISTRIBUTION | | | | | | | | | | | |
| NAME | QUANTITY | | USAGE | | COST | | LABOR | | RENTAL | | INTANGIBLES | | EXPENDIBLES | | MATERIAL | | SUBCONTRACT | |
	NUMBER	UNIT	NUMBER	UNIT	UNIT PRICE	TOTAL	U.P.	AMOUNT	U.P.	AMOUNT	U.P.	AMOUNT	U.P.	AMOUNT	U.P.	AMOUNT	U.P.	AMOUNT
Trench 1,500 LF																		
Machine w/Oper	1	Mch	10	Day	1,700.00	17,000			$1,700	17,000								

Assembly of equipment and facilities to make them ready to construct the project is collectively referred to as "construction plant" or simply "plant."

Mobilizing for a large project (e.g., the construction of a dam) can be a significant and costly undertaking that could include clearing, excavation, and grading; setting up a camp for workers and their families; and hauling in and setting up large pieces of equipment. Mobilizing for small jobs, while not of this scale, also requires careful thought and planning; mobilization is essentially a project within a project. Since you, the contractor, usually pay the costs associated with mobilization up front, careful planning is worthwhile.

All too often mobilization costs are overlooked when bidding a small job even though they may be significant. You should keep accurate records of mobilization costs so that you can recover them and also to assist with future estimates. Some firms consider mobilization to be an indirect expense, while others break mobilization down into parts (e.g., excavation, concrete, etc.) and allocate each part to the direct expenses of the task to which it is particularly involved, and still others keep mobilization as a separate aggregation of "mobilization expense" that is distributed during the summarization procedure. Regardless of how your accountant handles these costs, as the estimator you need to calculate them and include them in your final project cost estimate.

When preparing your mobilization estimate, do not forget moving costs, security, insurance, and if equipment is involved, unloading and reloading at your yard and at the jobsite and assembling equipment (unless considered separately). Other things to consider as you mobilize for your project include the following:

- Tool and plan storage. Set up a place in which to store tools and materials and read plans. If you are lucky, the owner may provide you with a garage or similar facility. If not, you'll have to construct a space. This may require leveling and construction. Depending on the job you may wish to set up some tools (e.g., a table saw, etc.), in which case you will need electrical power and perhaps heat. Make sure this area is secure and that you have enough copies of plans on site, including one set for each subcontractor.

- Security. Secure the work site and your tools and materials. This may entail constructing a fence and even hiring a night guard.

- Signage. Erect a sign with your company's name on it.

- Access. If you anticipate large deliveries (e.g., ready-mixed concrete or a semitrailer), make sure the trucks can get in and out of the site.

- Permits. Have a copy of the building permit available for when a building inspector arrives at your site; some localities require that building permits be visible to the public. Arrange for other permits as required. For example, you may need a permit to place a dumpster in a public area. Know the rules pertaining to the covering of dumpsters.

- Electricity. Arrange for temporary electric power if required. This may be from a utility company, or you may choose to use a portable generator.

- Heating. Arrange for temporary heat if required. Do not try to skimp on this only to learn later that your workers are not producing as you estimated. Workers produce better when they are not freezing. See Chapter 6, section 6.10 (Sizing Portable Heaters) for an example of how to size portable heaters.

- Parking. If parking is a problem, figure out where your workers will park so that you do not lose time the first day of work.

- If your work entails interior renovation that abuts space in which building occupants will be working you should familiarize yourself with the Sheet Metal and Air Conditioning Contractors National Association's (SMACNA) *IAQ Guidelines for Occupied*

Buildings under Construction.[3] Preparing your work site to ensure that you do not adversely affect occupants requires careful planning and may be costly; do not overlook this cost item. (On the positive side, informing clients that you abide with SMACNA's guidelines may help you sell yourself.)

PLANT OPERATIONS

Most large contractors consider plant-operating expense separately despite it being part of direct expenses because it has a form of its own. For them it is almost entirely the cost of operating construction equipment (e.g., the sum of fuel, lubricants, maintenance, labor, repair parts, ownership, operating labor, and abrasion). When estimating for equipment-intensive projects, the estimator must understand all factors. For example, the cost of parts used to overhaul a piece of equipment may be estimated separately or included under the category "normal repairs." Similarly, maintenance labor may include only labor applied to overhauls or may include greasers and so forth used for routine maintenance, or greasers may be included in operating labor. The point is that an estimator for a large contractor not only must take care to include all plant operations costs but also must not double count. It is somewhat easier for a small contractor, unless, of course, equipment costs are substantial. I shall assume this is not the case.

If you are a small contractor with a sizable equipment inventory, your office manager should keep an equipment ledger with a separate sheet or card for each unit. It would be to this record that the cost of each unit would be posted, and it is in the equipment ledger that control is kept of shifting combinations. For

---

3. SMACNA, *IAQ Guidelines for Occupied Buildings under Construction*, 2nd ed. (Chantilly, VA: SMACNA, 2007), chapter 3.

example, a costly diamond blade may be shifted back and forth between saws. In a well-kept system such a change will show on work orders. When the cost of the work order is posted, the necessary changes will also be made in the equipment ledger. At least once a year a physical inventory should be made and reconciled with the equipment ledger. Each significant piece of equipment should be given a job number or a company equipment number for identification.

Keeping a record of equipment costs is fairly simple. Whether owned or rented, all relevant costs are charged to a "cost of equipment operation" account (or something similar). These costs are, in turn, charged to the project accounts at some predetermined hourly rate. Under normal circumstances the amount charged to projects will exceed that charged to the cost of equipment operation account. The resulting balance is later used to absorb the cost of overhauls and other extraordinary repairs. From this you can see the important role your accountant, who usually establishes the hourly charge rate, plays in estimating.

If your firm's equipment costs are a substantial part of the cost of your work, it is important that equipment time is kept as carefully as labor costs because you will need these costs for billing, future estimates, making replacement decisions, and so forth. On the other hand, if your firm has little or no equipment aside from a pickup truck and several power-operated hand tools, you really do not need any special accounting for equipment costs. Your accountant can calculate their costs based on depreciation tables, taxes, insurance, fuel, and other maintenance costs. Just when to start tracking equipment costs is a management decision. Remember, however, that accurate unit costs are required for accurate estimating.

DIRECT EXPENSES

Direct expenses must be costed and tabulated systematically; they need to be classified in a logical and convenient form. By now your project will have been broken down twice: once for work analysis into elementary construction operations for cost determination and once into tasks for programming. It is practical to classify direct expenses by combining these two breakdowns.

Work analysis is done specifically to serve costing. Therefore the elementary construction operations should form the basis of your costing classifications. But the aggregate costs of the tasks used in programming are also needed for summarizing and presenting your estimate as a whole. Tasks are made up of one or more elementary construction operations. Therefore, you should do the following:

1. Use the task classification devised for programming as a primary classification.
2. Subdivide each task into its applicable elementary construction operations as you derived them during work analysis.

You should be aware of this compound classification as you do your takeoff during work analysis so that you can logically tabulate your results for costing.

Costing is a process of comparing work to be done with work that has already been done (on previous projects) and pricing it appropriately for the specific project. This comparison is made on the basis of costing items required in quantities of work, not on the basis of money, since prices change constantly. This is why you must be very leery when using estimating-guide prices and, in fact, even your own past pricing experience. You may find it convenient to make a distinction between costing data and pricing data. Costing data are used to designate cost data as described previously. Pricing data are applied to the costing data to obtain the cost of the project you are estimating. Pricing data should be current and local. For example, pricing data would consist of current labor rate schedules, current price lists, quotes, and so on.

INDIRECT AND OVERHEAD EXPENSES

As a rule, any expense that can be costed to a specific task should costed as such. But as already discussed it is not always practical to identify expenses with a specific project. These are called indirect expenses and overhead expenses.

There is fairly good agreement with regard to distinguishing between direct and indirect expenses. But there are borderline costs that are not always placed in the same category. Some variations are due to differences of opinion, and other times it simply costs more than it's worth to identify a specific cost to a specific job. This is where your accountant comes in. Resolve all issues with your accountant—preferably prior to starting your estimate or at least as the issues are identified.

5. COST DISTRIBUTION AND SUMMARIZATION

Until now costs and costing factors have been discussed as if they had no differentiating characteristics. This concept could be accepted if all we had to do was estimate a final project price. But your estimate is inextricably intertwined with cost keeping, cost control, and billing. Therefore all costs must be classified, not only as to tasks and their subdivisions, but also as to kinds of costs.

As already discussed, the basic subdivisions underlying such classifications are (1) staffing (labor), (2) equipment and equipment usage (rental), (3) intangible expenses (intangibles), (4) tangible items that are not incorporated into the work (expendables), (5) tangible items that are incorporated into the work (material), and (6) subcontract costs (subcontracts). Some systems may elaborate on these classifications, but I am unaware of any that violate them.

INDIRECT AND OVERHEAD COST DISTRIBUTION

Most small contractors use the sum of direct costs as the basis for distributing indirect and overhead costs. Others advocate distributing indirect and overhead costs on the basis of labor costs with the theory that this is more representative of the activities from which indirect expenses are derived. Examples of both methods are shown here.

Going back and forth between the written word and Figures 5.1, 5.2, 5.3, and 5.4, assume that a small contractor has four projects in progress during a given month and that the direct costs are as shown in Figure 5.1 and that the indirect costs are as shown in Figure 5.2; a pro rata share of these indirect costs must be added to each project.

To do this using the sum of direct costs as the basis of distribution, you would make the following calculations:

$$7188 \div 34{,}578 = 0.21$$

$$0.21 \times 8560 = 1779.$$

Using the sum of labor costs as the basis of distribution you would make these calculations:

$$3720 \div 17{,}910 = 0.21$$

$$0.21 \times 8560 = 1779.$$

In this case the result is the same.[1]

Figures 5.1 and 5.2 show the company's direct and indirect expenses. The amount of indirect expenses that would be distributed to each project are shown in Figures 5.3 and 5.4, depending on whether you are using total direct costs or total labor costs as your basis of distribution.

---

1. These calculations assume a small operation. There is no segregation of indirect costs between the home office and project or between overhead and indirect costs. As your operation grows your accounting system would become more sophisticated.

Figure 5.1. Company direct expenses, February XXXX

HOME BUILDING AND REPAIR INC.
DIRECT EXPENSES - FEBRUARY XXXX

Project	Labor	Materials	Equipment Rentals	Total
1	$3,720.00	$3,068.00	$400.00	$7,188.00
2	3,560.00	2,450.00	250.00	6,260.00
3	6,950.00	6,220.00	1,200.00	14,370.00
4	3,680.00	2,780.00	300.00	6,760.00
	$17,910.00	$14,518.00	$2,150.00	$34,578.00
			Check:	$34,578.00

Figure 5.2. Company indirect expenses, February XXXX

HOME BUILDING AND REPAIR INC.
INDIRECT EXPENSES - FEBRUARY XXXX

Office salaries	$1,460.00
Mortgage on office	3,078.00
Telephone	300.00
Internet, computer fees	125.00
Postage	97.00
Office supplies	220.00
Repairs & Maintenance	447.00
Small tools & supplies	520.00
Insurance	750.00
Taxes	590.00
Depreciation	730.00
Other	243.00
Total to be distributed:	$8,560.00

Figure 5.3. Distribution of indirect expenses

HOME BUILDING AND REPAIR INC.
DISTRIBUTION OF INDIRECT EXPENSES TO PROJECTS

Amount to be distributed - Based on total direct expenses $8,560

Project	Direct Costs	Distributed Amount
1	$7,188	$1,779
2	6,260	1,550
3	14,370	3,557
4	6,760	1,674
	$34,578	$8,560

Note: Based on total direct expenses.

Figure 5.4. Distribution of indirect expenses

HOME BUILDING AND REPAIR INC.
DISTRIBUTION OF INDIRECT EXPENSES TO PROJECTS

Amount to be distributed - Based on total labor expenses: $8,560

Project	Direct Costs	Distributed Amount
1	$7,188	$1,779
2	6,260	1,701
3	14,370	3,321
4	6,760	1,759
	$34,578	$8,560

Note: Based on total labor expenses.

PROVISIONAL RATE FOR ESTIMATING

This example is provided to give you an idea of how indirect and overhead costs are prorated to various projects. As already mentioned, the costs shown are incurred costs, whereas you are estimating a future project. Establishing a provisional, indirect, and overhead rate you will use for your estimate does this. This provisional rate is generally developed by computing the indirect and overhead rate of the previous year and adjusting it for known changes (e.g., increases in unemployment insurance).

Large contractors follow essentially the same principles, but they often reallocate the indirect and overhead costs that the home office has allocated to a project to individual bid items or major components.

SUMMARIZATION

As a small contractor you may be asked to provide a bid to a homeowner or someone who wants a clear, concise explanation of the price of your bid. In such a case you may wish to present your bid in essentially the order that you prepared your takeoff and collected your costs, showing what you wish to show or not show. For example, for home construction you may provide a number for each of the following if they apply:

1. Concrete
   a. Substructure
   b. Superstructure
   c. Finishes
2. Masonry
   a. Exterior
   b. Interior
3. Carpentry
   a. Rough
   b. Finished
4. Finishing trades (those that the general contractor would do)
5. Excavation
   a. Building
   b. Site

6. Site work
7. Subcontracts
   a. Electrical
   b. Heating, ventilating, and air conditioning (HVAC)
   c. Tile
   d. Other
8. Alternates

The price that you place next to each category would include your direct and indirect costs and any markup you might add. Or sometimes you might show only direct and indirect costs and show your markup as a separate line, thereby impressing your client how little money you make. Or you may wish to separate labor and materials. These are business decisions, not estimating decisions. You may then be asked to submit a lump-sum bid or a bid that shows costs on a unit-price basis. Bids that ask for unit prices usually specify the bid items. Regardless of how you present your estimate, you must first summarize your costs. This requires that you apportion indirect costs to your direct costs, as was discussed in the previous section.

You were given a simplistic—perhaps overly simplistic—example of how to do this. A more realistic example is provided in Chapter 6, section 6.11. Good, workable summary forms are shown within this detailed explanation on summarizing costs. In that case I show the summary using essentially the same tasks (slightly modified) that were used in the CPM example. You could just as easily list the categories you used to accumulate costs during your takeoff or the bid items requested by the owner.

Contractors, especially those that bid on government projects, have chafed at what they see as financing a project through the withholding of retainage. Some state legislatures have recognized this problem and are trying to help them out by putting limitations on the amount and length of time that retainage can be held. Nevertheless, many contractors and subcontractors have engaged in the guerilla warfare of front-end loading their bids by disproportionately pricing individual bid items to allow them to make up for the withholdings. From the contractor's perspective, this practice carries risks (1) of being disqualified; (2) of having the lower-priced-item quantities increased or the higher-priced-item quantities decreased, in which cases the contractor may suffer financially; and (3) of the owner, a public entity, having filed a false claims act (or equivalent), in which case the contractor might find himself in criminal court.

From the owner's perspective, accepting an unbalanced bid and paying out early may result in insufficient funds remaining to pay for the rest of the work and paying for necessary change orders (which can also lead to problems for an owner with a contractor's surety). When all is said and done, if you are inclined to play with your numbers, you might be better off in Las Vegas, Atlantic City, or a riverboat casino.

6. MISCELLANY

6.1 INSURANCES

There are several types of insurance that may be your (the contractor's) responsibility: workers' compensation, public liability and property damage, contingent liability, and hold harmless. Workers' compensation and public liability and property damage rates vary according to the trades involved and the state in which the work is being performed. States can vary by as much as 100%. Rates also vary based on the contractor's safety record, which is one reason it is important to have a good safety record. Standard rates are usually applied to contingent liability and hold harmless insurances.

Because rates vary widely, many contractors either carry these insurance items on the applicable trade sheets or use the labor totals from those trade sheets to compute a total insurance item for inclusion in overhead. Other contractors depend on their accountant to give estimators insurance costs as a percentage of payroll costs for particular types of jobs.

The insurance company auditors who visit a contractor's office to audit the payroll for an insurance breakdown are generally not technically trained construction people. They will often simply allocate wages into whatever classification might seem reasonable. Your accountant should work with the auditors to avoid an incorrect allocation.

6.2   SUBCONTRACTOR SELECTION

The work that your subcontractors do reflects directly on you. Indeed, their poor work will usually affect your reputation, not theirs. Therefore it is important to select outstanding subcontractors. Here is a method to help you with your selections.

Hold a preproposal meeting with the subcontractors that you invite to submit a proposal. At this meeting discuss the project in detail, including the following:

- The scope of work
- Performance milestones
- Inspection and acceptance criteria
- Evaluation criteria
- Any other material they will need to know to submit their bids to you (e.g., technical, financial, managerial, safety, and reporting requirements)

At this meeting inform them that you will expect them to complete the questionnaire that is shown in Figure 6.1. This questionnaire may seem too probing to some small subcontractors, and they may resist answering all your questions. Remember, however, on your project they will represent you; no one will remember their names. And since both you and they should hope for a continuing relationship, this information is important. You do not want a subcontractor going broke on one of your projects. Naturally, your firm and each project are unique, so you may want to alter the subcontractor qualification questionnaire that is presented (Fig. 6.1).

From the responses you receive to the questionnaire, you can form a short list of potential subcontractors. Using this short list, develop a decision matrix to help you choose the best firm for your project: follow the steps that are listed here,

Figure 6.1. Subcontractor qualification questionnaire

Home Building and Repairs Inc.
Subcontractor Qualification Questionnaire

1. What year was your company founded?
2. List your gross sales:
    a. Year 1
    b. Year 2
    c. Year 3
3. List projects currently under way, their expected completion date, and the value of each.
4. Have you ever filed for bankruptcy? If yes, explain the situation.
5. Provide the number of your full-time employees, by trade.
6. Provide a list of key and supervisory personnel, along with the years they have been with your firm.
7. If you are a union shop, list the name of the unit and the expiration date of your current agreement.
8. Provide three references, along with a description of the work done for them.
9. Do you currently have in-place contracts for supplies, materials, and equipment rentals? If yes, list your vendors and the nature of your agreements. If no, list three frequently used vendors.
10. Provide your safety record for the last 3 years. Explain your safety program in detail.
11. Provide financial reference(s) with contact name and phone number.
12. Provide insurance reference(s) with contact name and phone number.
13. Provide bonding reference(s) with contact name and phone number.
14. Describe in detail your firm's greatest strength.

and as you read, move between the written word and the decision matrix shown in Figure 6.2.

1. List those qualifications (your objectives) that you think are important.

2. Rate each qualification based on importance on a scale of 1 to 5 or 1 to 10. This is an important step because not all qualifications are equally important and should not be considered equally. You may consider some qualifications to be "musts," in which case any subcontractor that does not have these qualifications is automatically eliminated. For this reason, unless the qualification is irrevocably a "must," it is generally better to assign a maximum value to it, but not make it a "must." In this example nothing was classified as a "must."

3. Rate each subcontractor on a scale of one to five against the listed qualifications. Importantly, just as in boxing, one of the alternatives should be rated at the maximum: in this case, 5. There can only be one winner (5), but for the other alternatives, you may at times find two alternatives equally good, in which case it is OK to rate them the same.

4. Multiply the qualification's importance by the alternative's (subcontractor's) rating.

5. When all qualifications have been rated, sum the columns and get a score for each subcontractor (i.e., each alternative).

After you have a score for all subcontractors, you will have an *apparent* winner—here, subcontractor 2. There is, however, one more important step in the process. Ask yourself what can go wrong with this decision? Challenge your decision from the perspective of its downside. If there is no debilitating downside, you have a decision. In the example shown, the apparent winner was not the highest in three important categories. You must now decide whether you can bring subcontractor 2 up to your standards of quality and safety; if not, you may choose to select

Figure 6.2. Subcontractor selection

HOME BUILDING AND REPAIR, INC.
SUBCONTRACTOR SELECTION

Qualifications	Importance	Sub #1		Sub #2		Sub #3	
		Rating	Score	Rating	Score	Rating	Score
Construction Capability							
Construction Quality	5.0	5.0	25.0	4.0	20.0	3.5	17.5
Schedule Control	4.5	3.5	15.8	5.0	22.5	3.5	15.8
Construction Capability	5.0	4.0	20.0	4.0	20.0	5.0	25.0
Management Capability							
Ability to coordinate work	4.0	4.0	16.0	5.0	20.0	3.5	14.0
Safety program	4.5	5.0	22.5	4.5	20.3	2.0	9.0
Financial Condition	4.0	4.0	16.0	5.0	20.0	3.5	14.0
Reputation							
Financial references	4.5	4.0	18.0	5.0	22.5	2.5	11.3
Customer references	5.0	5.0	25.0	4.0	20.0	4.0	20.0
Supplier references	4.0	4.0	16.0	5.0	20.0	4.0	16.0
			174.3		185.3		142.5

Note: What can go wrong with subcontractor 2?
1. Construction quality is second to subcontractor 1.
2. Safety program is very good but second to subcontractor 1.
3. Customer references are second to subcontractor 1.

subcontractor 1 on the strength of the three categories for which he received a 5 rating. Also note that at this point pricing has not come into play. This, too, may alter your final decision.

6.3 ALTERATION WORK

Always separate alteration work from new construction because alteration work is almost always more costly than new work of the same type. For example, face brick for filling several isolated openings will cost more per brick to lay than face brick for an entirely new wall, and patching the wood base in an existing room costs more for labor in proportion to the amount of material used than setting a wood base in a new building.

The alteration work must be examined when you visit the site. Therefore the items involved should be clearly set out so that they may be identified and evaluated as you go through the building. The best way to set out such items is according to an orderly progression through the building room by room, always noting their locations.

Other considerations when dealing with alterations include the following:

- Architects do not always show alteration work in full detail. For example, they may simply note, "Remove partition." However, when you do your inspection you find that the partition is really a load-bearing wall for two floors.
- You may find that you will have to perform takeoff on some items during your inspection, measuring and describing the various items as you see them.
- The finish materials to be patched matched may be difficult to obtain.
- Demolition, removal of waste, and working conditions may be very difficult.

6.4   CONCRETE FORMS

The weight of concrete in a concrete form depends primarily on the density of the aggregate used. Concrete with dense aggregates may weigh as much as 160 pounds per cu. ft. (4320 pounds per cu. yd.) and sometimes more. Lightweight concrete may weigh as little as 75 pounds per cu. ft. (2025 pounds per cu. yd.). Most regular concrete weighs approximately 150 pounds per cu. ft. (4050 pounds per cu. yd.).

The pressure exerted by concrete on formwork varies with the following factors: rate of placement, temperature of the concrete, weight and density of the concrete, method of consolidating the concrete, and depth of concrete lifts. Make sure concrete forms are properly designed. You do not want to have to break up and haul away a day's pour.

6.5 HYDRAULIC AND MACHINEROOMLESS ELEVATORS

Only seldom are small contractors involved with elevators, and when they are, the elevators are installed by specialty contractors. The following two checklists (Figs. 6.3 and 6.4) are provided to help you inspect the most likely type of elevators to be installed in buildings of six stories or less. The checklists may also help you to speak with your subcontractor when you negotiate his contract.

6.6 VALUE ENGINEERING

Usually the purpose of value engineering (VE) is to save money and, at the same time, provide an equal or better value of goods and services. But sometimes it is to improve on some other established value. VE began at General Electric Co. during World War II when there were dire shortages of resources, and substitutes were needed. Surprisingly, many substitutes reduced costs, improved products, or both. What started as a need to find less expensive and completely satisfactory ways to meet the intent of the contract turned into a systematic, institutionalized process to accomplish this end. Now most large government contracts have a VE clause. Normally you would do VE after your estimate is completed, and you have a full understanding of the project.

On its surface VE provides you with a systematic approach to looking at what you are doing with an eye to cost cutting

Figure 6.3. Checklist: Hydraulic elevators

	Sat	Unsat	NA		Sat	Unsat	NA
Device Number				Numbering: Controllers,			
Rated Speed (FPM)				machine MG, disconnect switches			
Capacity (Pounds)				Code Data Plate			
Manufacturer							
Service Contractor							
Inside of Car	Sat	Unsat	NA	**Machine Room (Continued)**	Sat	Unsat	NA
Door reopening device				Hydraulic cylinder			
Stop switch				Pressure switch			
Operating control devices				Winding drum machine			
Car floor & landing sill				Hydraulic power unit			
Car lighting & receptacles							
Car emergency signal				**Top of Car**			
Door closing force				Stop switch			
Car door/gate				Car top light & receptacle			
Power closing of door/gate				Top of car operating device			
Power opening of door/gate				Clearance & refuge space			
Car vision panel/glass door				Normal terminal stopping device			
Car enclosure: width, depth				Emergency terminal speed			
Emergency exit				limiting device			
Ventilation				Anti-creep leveling device			
Signs/operating symbols				Top emergency exit			
Standby Power Operation				Floor & emergency ID			
Restricted openings				numbering			
Car ride				Hoistway construction			
				Hoistway smoke control			
Machine Room				Pipes, wiring & ducts			
Access to machinery/space				Windows, projections &			
Headroom				setbacks			
Lighting/receptacles				Hoistway clearances			
Housekeeping				Multiple hoistways			
Ventilation				Traveling cables &			
Fire extinguisher				junction boxes			
Pipes, wiring, ducts				Hoistway door & gate			
Exposed equipment/guards				equipment			
Disconnecting means				Car frame and stiles			
& control				Guide rails fastening			
Controller wiring, fuses				equipment			
& grounding				Governors, ropes			
Governor, over-speed				Suspension rope			
switch & seal				Speed test			
Control valve				Slack cable switch (roped			
Tanks				hydraulic)			
Flexible hydraulic hose				Counterweight			
& fittings				Door & gate equipment			
Supply line & shut-off				Car frame & stiles			
valve				Guide rails fastenings			
Relief valve				& equipment			
Top of Car (Continued)	Sat	Unsat	NA	**Pit**	Sat	Unsat	NA
Governor rope				Pit access, light, SW,			
Governor releasing carrier				receptacle & condition			
Wire rope fastening				Bottom clearance, runby &			
& equipment				minimum refuge space			
Suspension rope				Plunger & cylinder			
Top counterweight clearance				Normal terminal stopping			
Car, overhead				devices			
& deflector sheaves				Traveling cables			
Broken rope, chain or				Governor rope tension			

Figure 6.3. Checklist: Hydraulic elevators (*continued*)

tape switch			devices		
Counterweight & counter-weight buffer			Car & frame platform		
Counterweight safeties			Car safeties & guiding members		
Compensating ropes & chains			Car buffer		
Ascending car overspeed protection			Guiding members		
Unintended car motion			Supply piping		
			Firefighters' Service		
			Smoke detector recall		
Outside Hoistway			Phase I key switch operation		
Car platform guard			Phase II key switch operation		
Hoistway doors					
Vision panels			**ADA**		
Hoistway door locking device			Are there both visible and verbal or audible door/openings and floor indicators (one tone=up, two tones=down)?		
Access to hoistway					
Power closing of hoistway door					
Sequence operation					
Hoistway enclosure			Are the call buttons in the hallway no higher than 42 inches?		
Elevator parking device					
Emergency doors in blind hoistways			Do the controls inside the cab have raised and Braille lettering?		
Separate counterweight hoistway					
Standby power select switch/panel			Is the emergency intercom usable without voice communication?		
Inspection control			Can the lift be used without assistance?		
			Are there both visible and verbal or audible door opening/closing and floor		

			ADA (Continued)	Sat	Unsat	NA
			indicators (one tone=up, two tones=down)?			
			Are controls between 15 and 48 inches high (up to 54 inches if a side approach is possible)?			
			Are elevator cabs at least 48 by 48 inches? If not, can usability be demonstrated?			
			Do elevator doors provide clear opening at least 36 inches wide?			
			Are car controls 36 inches to 54 inches high?			

Figure 6.4. Checklist: Machineroomless elevators

Device Number				Numbering: Controllers,			
Rated Speed (FPM)				machine MG, disconnect switches			
Capacity (Pounds)				Code Data Plate			
Manufacturer							
Service Contractor							
Device Number							

Inside of Car	Sat	Unsat	NA	Top of Car (Continued)	Sat	Unsat	NA
Door reopening device				Gears and bearings			
Stop switch				Beltor chain drive machine			
Operating control devices				AC drive from DC sources			
Car floor and landing sill				Secondary and deflector			
Car lighting receptacles				Rope fastenings			
Car emergency signal				Terminal stopping devices			
Car door or gate				Governor, over speed switch			
Door closing force				Governor, over speed switch			
Power closing of doors or gates				Stop switch			
Power opening of doors or				Car top light and receptacle			
Car vision panels and glass				Top of car operating device			
Car enclosure: width				Top of car clearance & refuge			
Emergency exit				Car locking Device			
Ventilation				Locking Bracket			
Signs & operating device				Top counterweight clearance			
Rated load, platform area &				Normal terminal stopping			
Standby power operation				Final terminal stopping device			
Restricted opening of car door				Broke rope, chain, or tape			
Car Ride				Car leveling device			
Control Space				Crosshead data plate			
Access to control space				Top emergency exit			
Headroom				Counterweight and CTW buffer			
Lighting and receptacles				Counterweight safeties			
Enclosure of control space				Floor, and emergency ID#			
Housekeeping				Hoistway construction			
Ventilation				Hoistway smoke control			
Fire extinguisher				Pipes, wiring and ducts			
Pipes, wiring, and ducts				projections, and setbacks			
Guarding of exposed auxiliary				Hoistway clearances			
Numbering of elevators,				Multiple Hoistway			
Controller wiring, fuses,				Traveling cables, junctions			
Emergency break released				Hoistway door & gate			
				Car frame and stiles			
Top of Car				Guide rails fastening equipment			
Overhead beam and fastenings				Governor rope			
Drive machine brake				Governor releasing carrier			
Drive machine				Wire rope fastening and hitch			
Suspension rope							
				Pit			
Outside Hoistway				Pit access, light, SW, receptacle			
Car platform guard				Bottom clearance and runby			
Hoistway doors				Car and counterweight buffer			
Vision panels				Final & emergency terminal			
Hoistway door locking device				Normal terminal stopping			
Access to Hoistway				Traveling cables			
Power closing of Hoistway				Governor rope tension sheave			
Sequence operation				Comp chains, ropes and			
Hoist way enclosure				Car frame and platform			
Elevator parking device				Car safeties and guiding			
Emergency doors				Buffers & emergency terminal			
Separate counterweight				Car safeties and guiding			
				Smoke detector recall			
Signage				Firefighters' Service			
In lobby to show location of				Phase I key switch operation			
Control Space				Phase II key switch operation			

Figure 6.4. Checklist: Machineroomless elevators (*continued*)

Firefighters' Service	Sat	Unsat	NA	ADA (Continued)	Sat	Unsat	NA
Smoke detector recall				Are there both visible and verbal			
Phase I key switch				or audible door opening/			
operation				closing and floor			
Phase II key switch				indicators (one tone=up,			
operation				two tones=down)?			
				Are controls between 15 and 48			
ADA				inches high (up to 54			
Are there both visible				inches if a side approach			
and verbal or audible				is possible)?			
door/openings and floor				Are elevator cabs at least 48 by			
indicators				48 inches? If not, can			
(one tone=up,				usability be			
two tones=down)?				demonstrated?			
Are the call buttons in the				Do elevator doors provide clear			
hallway no higher				opening at least 36			
than 42 inches?				inches wide?			
Do the controls inside the cab				Are car controls 36 inches to 54			
have raised and Braille				inches high?			
lettering?							
Is the emergency intercom							
usable without voice							
communication?							
Can the lift be used without							
assistance?							

while maintaining or improving the value of your efforts. In addition, the application of VE principles improves teamwork and efficiency and forces your team to strive harder than ever for excellence. VE is functionally life-cycle oriented and draws creativity and innovation out of people.

When we speak of "best value," we are usually thinking of obtaining something—say, a car or facility—that meets our needs quality-wise at the lowest price over its useful life. Sometimes, however, value may mean something altogether different. For example, aesthetics and historical correctness may be more important than money to some owners. The VE approach can be used to improve on any definition of value you may wish to incorporate into your work.

On large projects value engineers are hired as consultants to take a carefully selected group through the formal VE process. Value engineers use a three-phase approach: (1) planning, (2) implementation, and (3) follow-up. They go on to break the implementation phase into five subphases: (1) information,

(2) creative, (3) analysis, (4) development, and (5) recommendation. Then the consultants depart, leaving the follow-up to others.

Unfortunately, as a first-line supervisor you will be given neither the resources nor the time to perform a formal and often long VE study. But you can, nevertheless, apply the principles of VE to many of your work assignments, which is why this section has been included here.

Planning Phase

Your team will be made up of the people you supervise, and if you are lucky, you may be able to obtain additional people who will bring special knowledge to your group. But during this phase, in which you will do the following, the onus rests on you:

- Establish a schedule. You may have to do some fast talking to get your boss to agree to take time to go through the process, at least the first time. (Once she sees what you have accomplished, the second time will be easier.) VE is not a day off; it is hard work.
- Plan how you are going to carry your team through the five subphases of the implementation phase. In planning for the information subphase, obtain as much information as possible: estimated project costs, operating costs, and constraints (local codes, ordinances, physical and environmental constraints, etc.).
- Make sure you have flip charts and markers, tape to hang the charts, paper and pencils, calculators, sufficient workspace, tables and chairs, and so forth. Blackboards and whiteboards are great if you have access to them.
- Distribute the preliminary data you collected so your team will be ready to work. Distribute only specific, relevant information that is free of biases and blame. If possible, array cost information

on a chart, graph, or both. When your team meets it will probably want to concentrate on the costly items. A 10% saving on a $1000 item is $100, while it is a savings of $10,000 is $1000. Yet the effort to cut 10% from the $10,000 item is probably no more than the effort needed to cut 10% from the $1000 item.

■ Establish the rules by which you will run your meeting. Prepare remarks to ensure that everyone will know what you hope to accomplish.

Implementation Phase

This phase will carry your group through a series of steps that almost always leads to a process or project of greater value.

Information Subphase

The information subphase consists of the following steps:

■ Provide the information you gathered during the planning phase. Have the group organize and convert your data into useful information. Validate the data (e.g., have the group check your cost estimates for reasonability). If appropriate, make a site visit.

■ Have your group establish its understanding of value. Usually value will be defined in term of money, but it might also be defined in terms of aesthetics and historical correctness, safety (use value), or something else. It does little good to spend time trying to save money when what you are trying to accomplish is to make your job safer.

■ Conduct a functional analysis. Function is the foundation on which VE is built. As with value, function is, to some extent, in the eyes of the beholder. Nevertheless, teams can usually agree on the functions for which expenditure is being made. For instance, roofs protect buildings from the elements and hard hats protect people from injury.

■   Attempt to state each functional requirement in two words,
     a verb and a noun. For example, the function of a sewer pipe
     can be described as "transport sewage." To help you define func-
     tion ask what, where, when, why, and how. The level of detail
     to which you and your team go is a judgment call but is nor-
     mally decided by the potential for improvement associated with
     the item or process. Here are additional examples of verb-noun
     combinations:

Verb	Noun
amplify	sound
collect	trash
emit	light
filter	v
impede	flow

■   Prepare functional analysis system technique (FAST) diagrams.
     Value engineers frequently draw FAST diagrams to better under-
     stand what is happening. By displaying functions in sequence and
     identifying missed, redundant, or unnecessary functions, FAST
     diagrams aid team members to visualize and understand their
     project better.

     FAST diagrams use a formalized set of rules. But by loosen-
     ing the rules (applying the principles, but with some personal
     preference adjustments to meet specific situations), the process
     seems to work better. The next diagram can be built upon with
     higher-order functions placed on the left and progressively lower
     order functions drawn on the right. Functions deemed to be of
     equal importance are stacked. FAST diagramming continues
     until the effort no longer seems worthwhile.

     Draw FAST diagrams from left to right. They are built
     by asking what, where, when, why, and how (but mostly how
     and why questions). Begin by placing the prime (basic) func-
     tion on the left. Then ask how (How do I [verb]?) and why

(Why do I [verb] [noun]?) questions about it. The answer to the why question should be placed to the left of the function. The answer to the how question should be placed to the right of the function. Repeat this process across the page. For instance, consider a school rooftop with a recreation area (Fig. 6.5).

Based on what you see on your FAST diagram, choose what you wish to discuss further. One way to do that is to compare cost to what you and your team deems it is worth. For example, if the design of the Heating, ventilating, and air conditioning (HVAC) support system is estimated to cost $10,000 and the people on your team say, "Hey, I know we can support that unit for under $7000," then you have a cost-to-worth ratio of 1.43:1 (10:7). As you look at the items, you will probably decide to look hardest at those items with the largest cost to worth ratio.

Creative Subphase

Having decided on what items you will study further, you now move into the creative subphase. Here your team lists recommendations on a blackboard or large sheets of paper. Usually ideas are shouted out to a recorder. To be fully effective, people must leave their egos behind; they can't worry about offending someone nor making an ass of themselves. In fact, during

Figure 6.5. Informal FAST diagram

Prime Function	Higher Order Functions	Lower Order Functions
Provide recreation area	Protect from falling objects	Allow view
Protect building	Avoid interference w/recreation area	Provide lighting
Support HVAC	Allow access to HVAC & antenna	Exhaust bad air

brainstorming two rules should be enforced: (1) no one may criticize any person or any idea, and (2) the merits of an idea may not be discussed; this will come later.

This process is known as brainstorming. Brainstorming is fun; it isn't every day that people are allowed to do unconstrained thinking. As the team leader, set a fast pace, calling on people when there is a lag. Quantity is desired without concern for quality. While remaining unconstrained, focus thoughts on ways to eliminate, simplify, modify, and combine ideas, while at the same time maintaining all the contractual requirements of the part of the project you are considering.

Analysis Phase

Now you have to do something with the long list just compiled. You should decide on the improvements you want to recommend to your boss. Here is where you narrow your list to a manageable size, examine the advantages and disadvantages of the remaining suggestions, make judgments, then reach conclusions.

This is the decision-making phase of the process. Some questions you should ask as you review the proposed ideas include the following:

- Will the idea save time or have other obvious cost advantages?
- Will function be affected?
- Will the idea affect other parts of the job, and how?
- Will excessive redesign be required?
- Are aesthetics adversely affected?
- Has the idea been used before?

As you ask your questions, try varying your technique. Since different people will respond differently to your leadership, consider using any of four basic types of questions: (1) overhead questions, addressed to the group as a whole; (2) directed questions that are aimed at a particular person; (3) reverse questions that, when you are asked a question, you turn back to the group; and (4) relay questions, which are relayed by you from one member to another.

At some point your group may come up with several ways of meeting the need that you are studying. For example, if you were handed a complex cribbing design to support an HVAC unit, your group might come up with a cheaper treated-wood cribbing and a way to tie the unit directly into the steel structure of the building.

Development Phase

Your team now carefully reviews its decisions. In some cases estimates may have to be confirmed, life-cycle costs calculated, and a recommendation prepared.

Recommendation Phase

This is when you sell your team's ideas to your boss. Without this, all your efforts will have been for naught. Therefore take time to prepare your recommendation. If you will be making an oral presentation rather than submitting a written report, make sure you rehearse. Refer to all previous sections on communication.

Follow-Up Phase

Nothing is more satisfying than to have your group's recommendation accepted and to be given the opportunity to actually make it happen.

6.7 NON-COMPUTER-BASED CRITICAL PATH METHOD

The basic Critical Path Method (CPM) is divided into three distinct phases:

- Phase I develops a "network model" of sequential relationships from the activities that make up the project. The proper visualization and construction of this network model is the most basic and important step and probably the most difficult part of CPM.
- Phase II develops information useful for control. From it a schedule may be presented or easily converted into a bar chart.

    When your CPM network is completed, you get a chain of operations deemed critical, hence the name "Critical Path." You can easily identify where a delay in the completion of a task will lengthen your project's time unless you do something, usually apply resources, to it. Without this knowledge you might be tempted to apply resources, as either overtime or extra equipment, to all activities, thereby wasting resources. It also identifies tasks that can be delayed without affecting project duration, thereby identifying resources for reallocation.
- Phase III seeks to determine the operating schedule that produces the least direct cost for a given project duration or to determine both the project duration and corresponding operation scheduling that produces the least overall project cost. Phase III is fairly difficult and time consuming to fully develop and is therefore seldom used for small projects. It is discussed only briefly.

At the end of this chapter there are hand-drawn diagrams that demonstrate the development of a CPM network for constructing a small house. (This example has been simplified, as I am introducing CPM and not home building. Please do not get caught up in the specific operations or their assigned times.).

During phase I, to develop a CPM schedule, you will need a well-defined collection of operations that, when completed, mark the end of the project.

First draw a rough diagram, as shown in Figure 6.6. At this time do not attempt to number the operations. Rather, represent each operation with a circle and a descriptive title. As you enter each operation ask yourself the following questions:

1. What operations must be completed to make this task possible?
2. Can this complete operation be done upon completion of the preceding operation?

Draw connecting lines between each operation to indicate which operation must directly precede or follow it. On this rough drawing the connecting lines may run in either direction. Therefore use arrowheads temporarily to indicate direction. When the rough network has been completed, determine the sequence step of each operation. The numbers that are on the lower left side of each circle in Figure 6.6 indicate sequence.

Draw the final network from the rough diagram: Figures 6.7, 6.8, and 6.9 take you through the steps. Place each operation indicated by a circle to the right of the operation preceding it. Try to place each operation horizontally at its proper sequence step. Try to place operations vertically so that connecting lines are represented clearly and, to the extent possible, as straight lines. Draw connecting lines, but arrowheads are no longer necessary. Next, number all operations in the same column or sequence step. Then go to the top of the next column and again number downward. Continue across your network.

Figure 6.6. Rough diagram

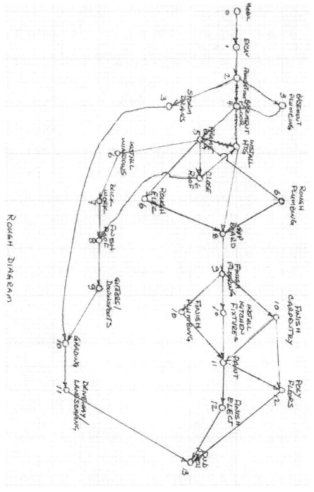

ROUGH DIAGRAM

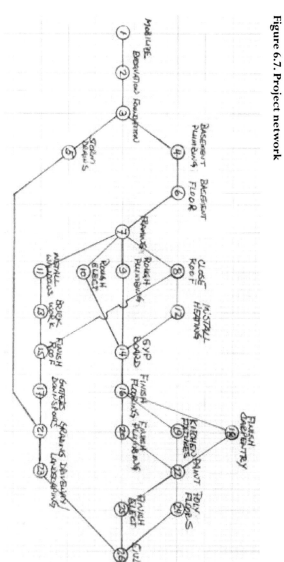

Figure 6.7. Project network

Figure 6.8. Circle notation

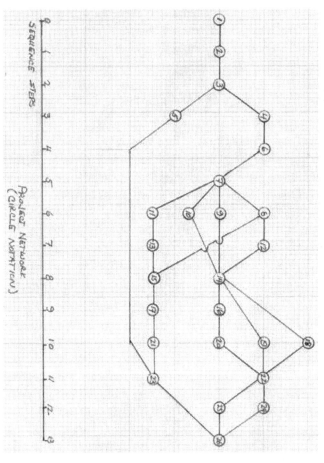

Figure 6.9. Critical path

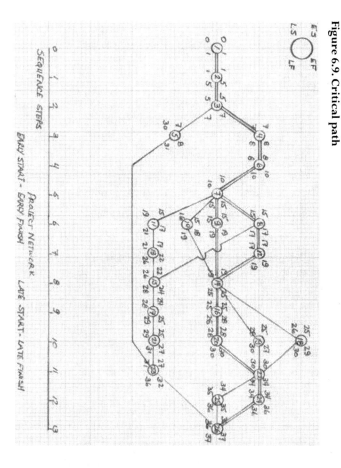

No operation can follow another operation of a higher number. The network phase (Figs. 6.7 and 6.8) is now completed. It should be clear that a network could be drawn for almost any process, including planning construction project, routing people, or even preparing your Thanksgiving dinner.

In phase II, you are now ready to prepare your schedule. Assuming the CPM technique will be applied through phases I and II only, the only additional needed information is a time estimate for each task. In estimating the time to accomplish each task, you need to develop a practical method of accomplishment. This consists of the following steps:

- Determine a means of access.
- Determine the tools, equipment, and plant needed.
- Determine crew size and rate of production.
- Determine the time to accomplish all aspects of the task, including mobilization, accomplishment, and demobilization, if necessary.

Usually these estimates would be based on your experience. When you lack experience you may wish to access any number of estimating guides that are available. Whole days are used typically for construction project CPMs. Once estimated times are available for each task, the computation work of phase II can be performed on either a worksheet or the network diagram. Most people work directly on small network diagrams; however, as some additional information can be obtained using a spreadsheet, this is also shown and explained.

First, work with the diagram only. As you work through the network, write the earliest start time of each operation at the upper left-hand side and the earliest finish time of each operation at the upper right-hand side of each operation. If an operation cannot begin until two or more previous operations must be completed, this operation cannot start until the latest of all the previous operations have been completed.

Next, determine the latest start and finish times for each operation. (Show the latest finish times at the lower right-hand side and the latest start times at the lower left side of each operation.) To do this work backward, with the starting point at the earliest finish time for the last operation. Then the latest start time is obtained by subtracting the operation duration time. Go back to Figure 6.9 (Critical Path) and follow the numbers through.

In this example, assuming that you have estimated durations properly, you could with confidence tell your client that you will be completed and off the site in 37 workdays. (This network does not factor in weekends and bad weather.)

You now know both the earliest and latest possible finish times for each operation and can determine what are known as "total floats." Float is a measure of the leeway available for an operation. The total float figure gives you the time by which the finish time of an operation can exceed the earliest possible finish time without affecting the duration of your overall project. Zero total float tells you that an operation has no leeway and, therefore, is an operation that establishes the duration of your project. If its finish time is moved back, the project will be delayed by an equal amount of time. Operations with no total float are labeled "critical operations." There must be at least one chain of critical operations running from start to finish of each project. This is about as far as you can go using the network diagram only.

Now let's look at both the network and the four spreadsheets: Figures 6.10, 6.11, 6.12, and 6.13. You need to work with both the network and the spreadsheets at the same time. With

Figure 6.10. Determine critical tasks, time boundaries, and floats, step 1

HOME BUILDING AND REPAIR INC.
DETERMINATION OF CRITICAL TASKS, TIME BOUNDARIES, AND FLOATS

Task	Est. Time	Start Earliest	Start Latest	Finish Earliest	Finish Latest	Float Total	Float Free	Crit. Op.
1	1	0		1				
2	4	1		5				
3	2	5		7				
4	1	7		8				
5	1	7		8				
6	2	8		10				
7	5	10		15				
8	2	15		17				
9	4	15		19				
10	3	15		18				
11	2	15		17				
12	2	17		19				
13	5	17		22				
14	6	19		25				
15	2	22		24				
16	3	25		28				
17	1	24		25				
18	4	25		29				
19	2	25		27				
20	2	28		30				
21	2	25		27				
22	4	30		34				
23	5	27		32				
24	2	34		36				
25	1	34		35				
26	1	36		37				

them you can obtain additional scheduling information known as "free float." The free float of an operation is the difference in time between its earliest finish date and the earliest of the "earliest start" times of all its directly following operations. The difference between total float and free float is called "interfering float" to indicate that, while operation completion in this time range does not affect project completion time, it does affect some subsequent operations by decreasing their float. For example, look at task 5. It has a total float of 23 days. This means that you can delay the start of this task 23 days and not affect project completion. But it has a free float of 19 days. This means that if you delay task 5 more

Figure 6.11. Determine critical tasks, time boundaries, and floats, step 2

HOME BUILDING AND REPAIR INC.
DETERMINATION OF CRITICAL TASKS, TIME BOUNDARIES, AND FLOATS

Task	Est. Time	Start		Finish		Float		Crit.
		Earliest	Latest	Earliest	Latest	Total	Free	Op.
1	1		0		1			
2	4		1		5			
3	2		5		7			
4	1		7		8			
5	1		30		31			
6	2		8		10			
7	5		10		15			
8	2		15		17			
9	4		15		19			
10	3		16		19			
11	2		19		21			
12	2		17		19			
13	5		21		26			
14	6		19		25			
15	2		26		28			
16	3		25		28			
17	1		28		29			
18	4		26		30			
19	2		28		30			
20	2		28		30			
21	2		29		31			
22	4		30		34			
23	5		31		36			
24	2		34		36			
25	1		35		36			
26	1		36		37			

than 19 days, while not necessarily affecting the project's completion time, you will affect the earliest start time of task 23.

The concepts of total float and free float can be seen graphically through use of a common bar chart. Figure 6.14 shows a bar chart for the same task breakdown as shown on the previously discussed network. The solid black bars show task durations and the earliest start and finish dates. Standing alone they represent a conventional bar chart. Bars that are not extended represent critical tasks. The full extension of the bars represent total float; however, these extensions are further shown as interfering float and free float. When there is zero free float, interfering float equals

Figure 6.12. Determine critical tasks, time boundaries, and floats, step 3

HOME BUILDING AND REPAIR INC.
DETERMINATION OF CRITICAL TASKS, TIME BOUNDARIES, AND FLOATS

Task	Est. Time	Start Earliest	Start Latest	Finish Earliest	Finish Latest	Float Total	Float Free	Crit. Op.
1				1	1	0		√
2				5	5	0		√
3				7	7	0		√
4				8	8	0		√
5				8	31	23		
6				10	10	0		√
7				15	15	0		√
8				17	17	0		√
9				19	19	0		√
10				18	19	1		
11				17	21	4		
12				19	19	0		√
13				22	26	4		
14				25	25	0		√
15				24	28	4		
16				28	28	0		√
17				25	29	4		
18				29	30	1		
19				27	30	3		
20				30	30	0		√
21				27	31	4		
22				34	34	0		√
23				32	36	4		
24				36	36	0		√
25				35	36	1		
26				37	37	0		√

total float. This bar chart can be constructed to show calendar days to provide space for actual progress of tasks, and to show percentage completion estimates, all common uses of a bar chart. You may also conveniently use the bar chart by "sliding" tasks within the free float zone without affecting the project and within the interfering float zone recognizing that you are affecting following task float times.

In summary, phase II provides information for you to both plan and control your project. It may be applied to data from ordinary estimating procedures as long as you use the same breakdown as in phase I. Phase II calculations are purely

Figure 6.13. Determine critical tasks, time boundaries, and floats, step 4

HOME BUILDING AND REPAIR INC.
DETERMINATION OF CRITICAL TASKS, TIME BOUNDARIES, AND FLOATS

Task	Est. Time	Start Earliest	Start Latest	Finish Earliest	Finish Latest	Float Total	Float Free	Crit. Op.
1		0		1			0	
2		1		5			0	
3		5		7			0	
4		7		8			0	
5		7		8			19	
6		8		10			0	
7		10		15			0	
8		15		17			0	
9		15		19			0	
10		15		18			1	
11		15		17			0	
12		17		19			0	
13		17		22			0	
14		19		25			0	
15		22		24			0	
16		25		28			0	
17		24		25			0	
18		25		29			1	
19		25		27			3	
20		28		30			0	
21		25		27			0	
22		30		34			0	
23		27		32			4	
24		34		36			0	
25		34		35			0	
26		36		37			0	

mechanical. If you use a computer with a spreadsheet program (e.g., Excel), you can obtain the results quickly. But for small jobs I find working through the network by hand helps me to better get my head into the job; you must decide for yourself.

Phase I of CPM provides a graphic network of the tasks that make up your project. Phase II provides scheduling information. But neither phase I nor phase II take cost into account. Phase III does this. Phase III provides you with the information to plot a project time versus cost curve, as well as points on the curve. The points on the curve are the lowest total direct costs at the corresponding project duration times.

Figure 6.14. Bar chart construction schedule

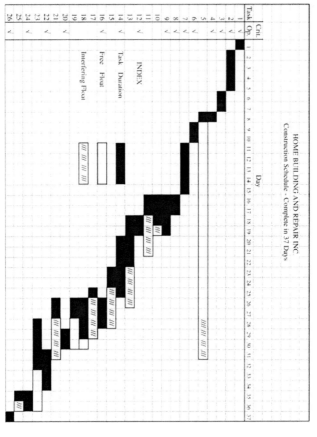

Since phase III deals with time-cost variations, a starting point is required. One point would be the "least direct cost" starting point. This point assumes that you based your phase II schedule on accomplishing each task at its lowest possible cost. This will give you the "least direct cost" starting point.

Another starting point could be determined by estimating that each task is accomplished as quickly as possible. This would be the most costly way of doing the project. If you then draw a network for these estimated task times, you would have the highest cost for the project; this would be called an "all crash" network. However, because not all tasks would be critical, you could ease up on some tasks so as not to lengthen the shortest possible project duration but still save some money.

You should now see where this is leading. If your client has established a time requirement that is shorter than the low-cost time initially established by you, you can through an iterative process determine the lowest possible cost to accomplish the project in the required time. The first step would be to lengthen any noncritical tasks. Then if you are below the client's required end date, you could lengthen a critical item that saves the most money yet not bring you past the client's deadline. Using this iterative process you can arrive at the best possible solution. A mechanical process for working through this process and obtaining the best possible solution has been established.[1] It is quite laborious and requires numerous estimates for each task time selected.

1. If you are interested in the nitty-gritty of phase 3, you should obtain a copy of *A Non-Computer Approach to the Critical Path Method for the Construction Industry*, 2nd edition, by John W. Fondahl (technical report no. 9, November 1961; rev. 1962). This book was prepared under research contract NBy-17798, Bureau of Yards and Docks, US Navy, and was distributed by the Construction Institute of Stanford University. It serves as the prime reference for this section on CPM and is the foundation on which computerized approaches to CPM sit.

There are two popular forms of network analysis in construction management practice: activity on the node and activity on the arrow. The activity on the node is also referred to as a precedence diagram. Each of these approaches offers virtually the same information, and whether to use one or the other is largely a matter of preference.

6.8   LINE-OF-BALANCE SCHEDULING

Some of your projects may include activities that are repetitive (e.g., a multistory building with similar floors). These repetitive activities can be modeled in CPM, but they can be difficult to visualize. The Line-of-Balance (LOB) schedule is a graphical technique that can be used in conjunction with CPM and may help you better visualize how work crews interact. As with the previous CPM drawings, I have done these LOB drawings by hand.

Follow along with the diagrams (Figs. 6.15, 6.16, 6.17, and 6.18) that are at the end of this section. On these drawings the x-axis represents the project's timeline and the y-axis identifies the work areas that in the example are the repetitive floors of a multistory building.

As crews arrive on the jobsite, they begin at the first floor and move through the project. Activity A has a total duration of 15 weeks. Assuming the crew spreads across the entire floor and that it can continue to work without interruption, the graph shows that activity A takes 15 weeks to complete (Fig. 6.15).

Activity B requires only 1 week to complete, but it cannot begin until activity A is completed on the first floor. If we want to start activity B as soon as possible, it could start on floor 1 on Monday morning of week 4 (Fig. 6.16). As we continue with activity B, we see that we cannot start on floor 2 until Monday

Figure 6.15. Activity A progress chart

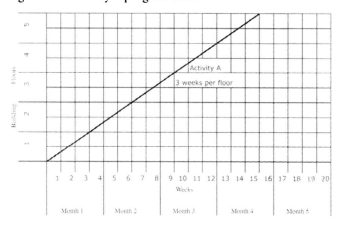

Figure 6.16. Earliest start of activity B, floor 1

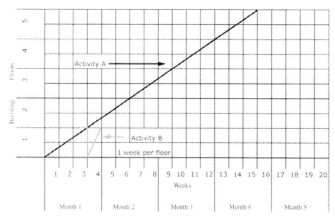

Figure 6.17. Earliest start of activity B, all floors (requires crew down time)

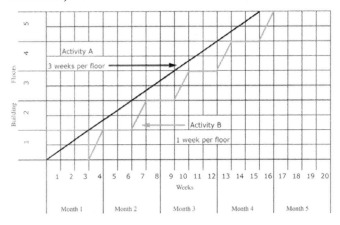

Figure 6.18. Late start of activity B, all floors (continuous operation)

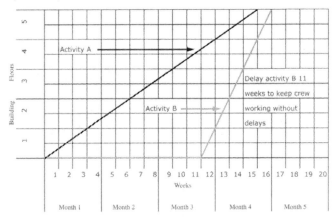

morning of week 7, following completion of activity A on this floor. This means that the crew for activity B would have to be idle or assigned elsewhere. As we follow through with activity B, we see that the crew cannot work continuously up the building. This starting and stopping usually results in inefficiency. Figure 6.17 is a visual presentation of this starting and stopping.

Figure 6.18 shows that if you should delay starting activity B for 11 weeks you could keep the activity B crew working straight through. This, of course, assumes that delaying activity B does not delay another activity.

As mentioned previously, this is merely an introduction to LOB scheduling. Research is ongoing. You can expect many refinements to come from both industry and academia in the near future.

6.9 Q. SCHEDULING

Q. Scheduling is a new technique that is gaining popularity within the construction industry. It is a scheduling technique that reveals a relationship between the sequence of doing a job and the cost incurred. The Q. Schedule is similar to the LOB schedule, but it allows for a varying volume of repetitive activities at different segments or locations of the construction project. The following example briefly explains the technique.

Assume a project site has three buildings, A, B, and C, with the quantities represented in Figure 6.19. If you want to ensure a continuous flow of work for each activity with no idle times for any crew and cannot allow more than one activity to take place at any one location, then there are six possible arrangements for doing the job (see Figs. 6.20–6.25):

Figure 6.19. Daily output quantities per building

HOME BUILDING AND REPAIR, INC.
THREE BUILDING PROJECT

Activity	Unit	Average Daily Output	Building A	Building B	Building C
Excavation	CY	25	75	25	50
Foundations	CY	20	40	20	20
Backfilling	CY	35	35	70	70

Figure 6.20. Alternative 1 gives a total duration of 10 working days.

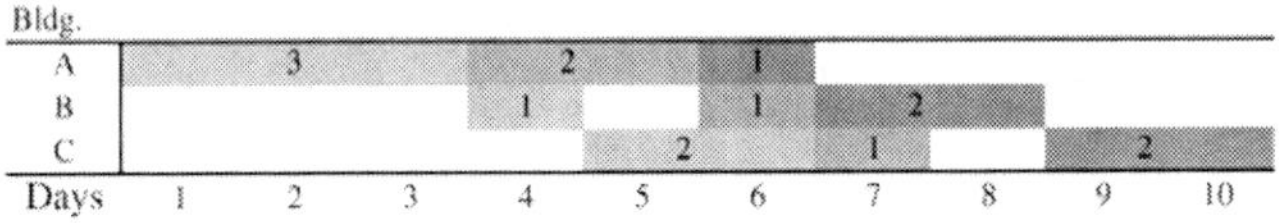

Figure 6.21. Alternative 2 gives a total duration of 10 working days.

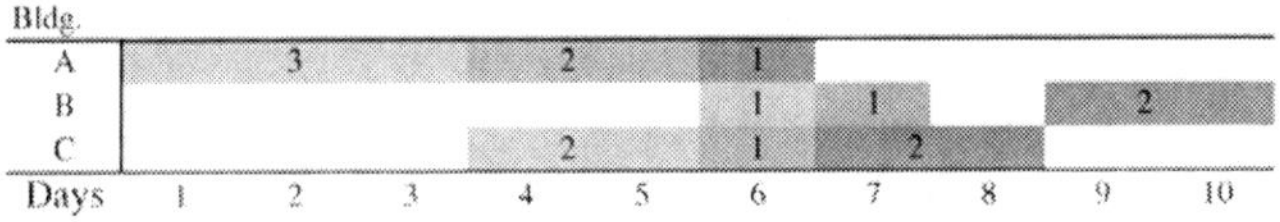

Figure 6.22. Alternative 3 gives a total duration of 9 working days.

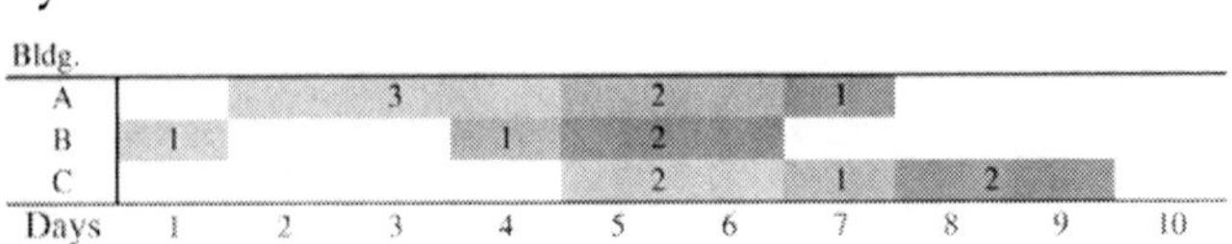

Figure 6.23. Alternative 4 gives a total duration of 10 working days.

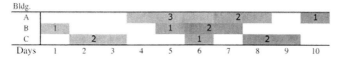

Figure 6.24. Alternative 5 gives a total duration of 10 working days.

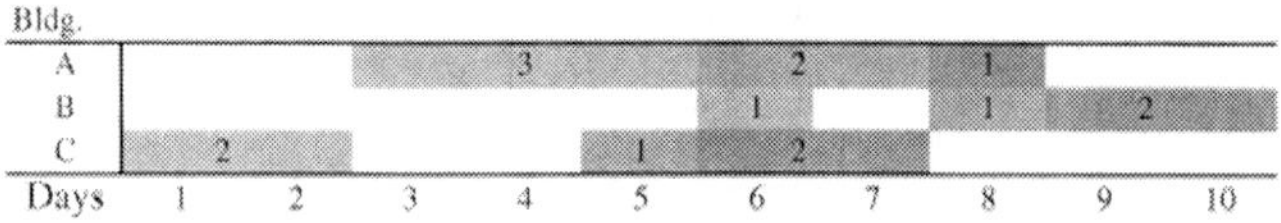

Figure 6.25. Alternative 6 gives a total duration of 10 working days.

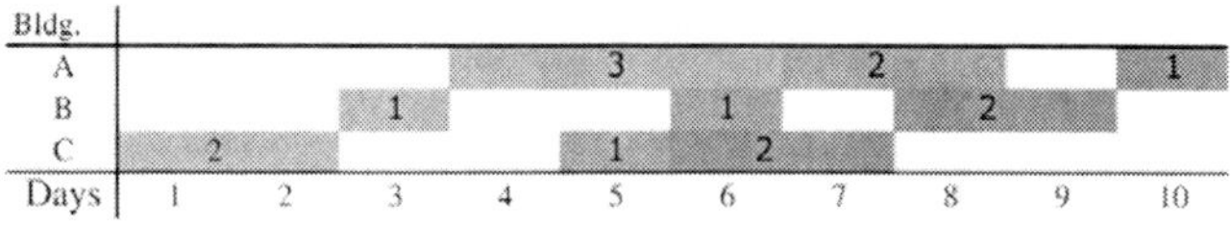

- A → B → C
- A → C → B
- B → A → C
- B → C → A
- C → A → B
- C → B → A

The examples show that alternative 3 produces a saving of 1 day (10%) relative to other alternatives assuming output rates are kept same. Q. Scheduling software would pick up alternative 3 as the most economic sequence and would consider other alternatives as having additional cost. Note, however, that a construction project having four locations would have 24 possible alternatives, and one with five would have 120 alternatives, so it is really not as easy as this simplified example might suggest.

6.10 SIZING PORTABLE HEATERS

Compact, wheeled portable heaters are typically used in residential construction. They range from in size from 25,000 to around 200,000 Btus and vary in cost depending on size, features, and quality. Because you may require different sized heaters across the life of a single job and because you may need them infrequently, they are a popular rental item, unless, of course, you regularly work through cold winter weather. You can generally receive good advice as to the proper heater for your job from the rental company.

Heaters are rated on the amount of Btus they produce per hour. When selecting a heater, make sure that it is large enough to replace the heat lost through the envelope of your work area. The amount of heat lost depends on how well the area is sealed and insulated. You can follow these steps to estimate your heating requirement:[2]

1. Determine the surface area of your floor, walls, and ceiling in sq. ft.:

    $$(2 \times length \times width) + (2 \times length \times height) +$$
    $$(2 \times width \times height) = surface\ area.$$

---

2. This method for estimating heater size was found in the McMaster-Carr catalog 116. See the McMaster-Carr website: http://www.mcmaster.com.

2. Estimate your heat-loss factor by choosing the description that best fits your work site:
   - Very well sealed and insulated = 0.25
   - Well sealed but not insulated = 0.75
   - Not well sealed or insulated = 1.25
3. Decide how much you want the temperature to rise in degrees Fahrenheit. If the area is unheated, this would be the difference between the outside temperature and your desired temperature. If adding to an existing temperature, use the difference between your current temperature and your desired temperature.
4. Multiply the results from steps 1–3 to obtain the Btu per hour you require. You may find that you will need several heaters.
5. If you are sizing for an electric heater by watts, multiply the result in step 4 by 0.293.

As an example, suppose you are adding on a 20 ft. L × 18 ft. W × 14 ft. H space to a house. The space will be over a crawl space. The outside temperature is around 25°F and you would like your workers to be at least 60°F.

1. Your surface area is

$$(2 \times 20 \times 18) + (2 \times 20 \times 14) +$$
$$(2 \times 18 \times 14) = 1784 \text{ sq. ft.}$$

2. Your heat factor will vary as the job progresses, but initially it will be 1.25, as the space will not be insulated or sealed.
3. You need a 35°F temperature difference.
4. Multiplying the results determines your heat requirement:

$$1784 \times 1.25 \times 35 = 78,050 \text{ Btu/hr.}$$

You would probably choose two heaters in the area of 40,000 Btu/hr each.

6.11 SUMMARIZING YOUR BID

A system of arithmetical proportioning is required to allocate indirect expenses to the various bid items. Most contractors allocate indirect costs based on the ratio of the direct costs of the bid item to the total direct costs. Another method is to allocate indirect costs in proportion to the labor cost of each bid item because direct costs are more representative of the activity from which indirect expenses are derived. Figure 6.26 is based on distributing indirect costs based only on direct labor, and Figure 6.27 is based on distributing indirect costs based on total direct costs.

The most conspicuous feature of this summary is that it identifies all costs in some way: individual tasks, bid items, or the categories used in doing a takeoff. This means that indirect expenses must be distributed among the items on some rational basis. The summary form is filled out as follows:

1. Under the "name" column enter the categories that you intend to use to summarize your bid. In the example these are tasks that must be accomplished. Under "quantity," enter the numbers and units as appropriate. Some owners provide this information, in which case, if you determine different amounts, you will have to adjust your pricing accordingly. Similarly, if you do your takeoff based on different units, you will have to adjust to the owner's wishes.

2. Enter the direct expenses (i.e., labor, rental, intangibles, material, and subcontract) that you calculated by doing your takeoff. Total each column of direct expenses.

3. Enter the total of indirect expenses that you will assign to this project. This is entered at the bottom of the "indirect" column. In the examples it is based on 25% of direct expenses, a figure that would be based on the company's experience and assigned to the estimator ($292,137 \times 0.25 = \$73,034$).

Figure 6.26. Distribution of indirect costs based on direct labor costs

HOME BUILDING AND REPAIR, INC.

SUMMARY SHEET

PROJECT XX-XX Oak Drive

EST. BY KFS

SHEET NO. 1 OF 1 SHEETS

BID ITEM			COST						COST DISTRIBUTION												BID			
	QUANTITY		TOTAL		INDIRECT		DIRECT		LABOR		RENTAL		INTANGIBLES		EXPENDIBLES		MATERIAL		SUBCONTRACT		TENTATIVE		FINAL	
NAME	NUMBER	UNIT	U.P.	AMOUNT	U.P.	AMOUNT	U.P.	AMOUNT	U.P.	AMOUNT	U.P.	AMOUNT	U.P.	AMOUNT	U.P.	AMOUNT	U.P.	AMOUNT	U.P.	AMOUNT	U.P.	AMOUNT	U.P.	AMOUNT
Grading	7,000	SF	0.17	1,200			0.17	1,200											1,200.00	1,200				
Excavation	60	CY	171.17	10,270	15.93	956	155.26	9,316	23.40	1,404	131.73	7,904			0.13	8								
Foundation	12	CY	367.19	4,406	110.60	1,327	256.56	3,079	162.67	1,952					22.92	275	71.00	852						
Storm Drains	125	LF	78.09	9,762	3.11	389	74.98	9,372	[illegible]	972	16.00	2,000			0.00		54.40	6,800						
Basement Floor	30	CY	106.28	3,188	14.28	428	92.00	2,760	23.00	690					0.00		71.00	2,130						
Rough Carpentry	127	MBF	1,298.38	164,899	331.81	42,140	966.54	122,750	488.00	61,976					3.54	450	475.00	60,325						
Finish Carpentry	1	LS	59,053.46	59,053	16,998.46	16,998	42,055.00	42,055	25,000.00	25,000					55.00	55	17,000.00	17,000						
Brickwork	1,280	SF	18.75	23,994	5.05	6,464	13.75	17,530	7.66	9,805					0.02	23	6.25	8,000						
Gypsum Board	3,100	SF	2.76	8,560	1.03	3,194	1.73	5,365	1.51	4,677					0.03	90	0.20	620						
Kitchen Fixtures	1	LS	10,130.00	10,130			10,130.00	10,130							0.00		10,130.00	10,130						
Paint	3,900	SF	0.97	3,789	0.29	1,139	0.67	2,650	0.43	1,675					0.02	95	0.22	880						
Driveway & Landscape	1	LS	17,500.00	17,500			17,500.00	17,500											17,500.00	17,500				
Poly Floors	1	LS	1,750.00	1,750			1,750.00	1,750											1,750.00	1,750				
Plumbing	1	LS	3,750.00	3,750			3,750.00	3,750											3,750.00	3,750				
Electric	1	LS	22,800.00	22,800			22,800.00	22,800											22,800.00	22,800				
HVAC	1	LS	20,150.00	20,150			20,150.00	20,150											20,150.00	20,150				
Total Direct				292,137				292,137		107,413		9,904				955		106,715		67,150				
Indirect				73,034		73,034																		
Total Project Cost				365,172																				
Indirect apportioned by cost category						73,034				35,913		3,127		2,000		301		33,693						

Note 1: Distribution Factor = Indirect Costs/Direct Labor. 0.25 X 292,137 = 73,034.31 73,034/107,413 = 0.68

Note 2: Here indirect costs are distributed based on total labor costs.

For example: Total indirect costs to be distributed to Excavation: 0.68 X 1,404 = 955.

Note 3: $2,000 of indirect expenses distributed to Intangibles based on experience.

Note 4: To distribute indirect expenses to cost categories first sum Labor, Rental, Expendables and Material. This equals $224,987.

Then divide 71,034 (Intangibles excluded) by 224,987 to get 0.325, your distribution factor.

Using Labor as an example, 107,413 X 0.325 = 35,913.

Figure 6.27. Distribution of indirect costs based on direct project costs

HOME BUILDING AND REPAIR, INC.

SUMMARY SHEET

PROJECT XX-XX Oak Drive

EST. BY KFS

SHEET NO. 1 OF 1 SHEETS

BID ITEM			COST						COST DISTRIBUTION				
NAME	QUANTITY		TOTAL		INDIRECT		DIRECT		LABOR		RENTAL		INTANGIBLES
	NUMBER	UNIT	U/P	AMOUNT	U/P	AMOUNT	U/P	AMOUNT	U/P	AMOUNT	U/P	AMOUNT	U/P
Grading	7,000	SF	0.21	1,500		300	0.17	1,200					
Excavation	90	CY	166.07	11,844	38.83	2,329	135.26	9,516	25.40	1,604	131.75	7,906	
Foundation	12	CY	320.73	3,849	64.15	770	256.58	3,079	162.67	1,952			
Storm Drains	125	LF	93.72	11,715	18.74	2,343	74.98	9,372	4.58	573	16.00	2,000	
Basement Floor	30	CY	115.00	3,450	23.00	690	92.00	2,760	21.00	630			
Rough Carpentry	127	MBF	1,208.18	153,439	241.64	30,688	966.54	122,751	468.04	61,976			
Finish Carpentry	1	LS	52,568.75	52,569	10,513.75	10,514	42,055.00	42,055	25,000.00	25,000			
Brickwork	3,200	SF	17.12	21,912	3.42	4,382	13.70	17,530	3.43	9,585			
Gypsum Board	3,100	SF	2.17	6,729	0.43	1,344	1.73	5,375	1.53	4,697			
Kitchen Fixtures	1	LS	12,650.00	12,650		2,530	10,120.00	10,120					
Paint	3,000	SF	0.84	3,288	0.17	658	0.67	2,630	0.43	1,675			
Driveway & Landscape	1	LS	21,875.00	21,875		4,375	17,500.00	17,500					
Poly Floors	1	LS	2,187.50	2,188		438	1,750.00	1,750					
Plumbing	1	LS	4,687.50	4,688		938	3,750.00	3,750					
Electric	1	LS	28,500.00	28,500		5,700	22,800.00	22,800					
HVAC	1	LS	25,187.50	25,188		5,038	20,150.00	20,150					
Total Direct				292,137				292,137		107,413		9,906	
Indirect				73,034		73,034							
Total Project Cost				365,172		73,734							
Indirect apportioned by cost category						73,034				33,953		3,127	

NAME	INTANGIBLES AMOUNT	EXPENDABLES U/P	EXPENDABLES AMOUNT	MATERIAL U/P	MATERIAL AMOUNT	SUBCONTRACT U/P	SUBCONTRACT AMOUNT	BID TENTATIVE U/P	BID TENTATIVE AMOUNT	BID FINAL U/P	BID FINAL AMOUNT
Grading						1,200.00	1,200				
Excavation		0.15	8								
Foundation		22.92	275	71.00	852						
Storm Drains		0.00		54.40	6,800						
Basement Floor		0.00		71.00	2,130						
Rough Carpentry		3.54	450	475.00	60,325						
Finish Carpentry		55.00	55	17,000.00	17,000						
Brickwork		0.02	73	2.25	8,000						
Gypsum Board		0.03	80	0.20	626						
Kitchen Fixtures		0.00		10,120.00	10,120						
Paint		0.02	95	0.22	860						
Driveway & Landscape						17,500.00	17,500				
Poly Floors						1,750.00	1,750				
Plumbing						3,750.00	3,750				
Electric						22,800.00	22,800				
HVAC						20,150.00	20,150				
Total Direct			955		106,715		67,150				
Indirect											
Total Project Cost											
Indirect apportioned by cost category	2,000		301		33,663						

Note 1: Distribution Factor = Indirect Costs/Direct Labor Priced at 25%

Note 2: Here indirect costs are distributed based on total estimated direct project total estimated project costs.

For example: Total indirect costs to be distributed to Excavation: 0.25 X 9,516 = 2,329

Note 3: $2,000 of indirect expenses distributed to Intangibles based on experience.

Note 4: To distribute indirect expenses to cost categories first sum Labor, Rental, Expendables and Material. This equals $224,987

Then divide 71,034 (Intangibles excluded) by 224,987 to get 0.325, your distribution factor

Using Labor as an example: 107,413 X 0.325 = 33,913

Note that of the $73,034 of indirect costs that will be assigned to this project, I have listed $2000 under intangibles. As with the 25%, this assignment was based on experience.

In the examples given, mobilization was considered an indirect expense. But as mentioned in Chapter 1, there are other ways to distribute mobilization expenses, including these three common approaches:

a. Treat mobilization as an indirect expense.

b. Charge mobilization through a suspense procedure to direct expense.

c. Separately aggregate from both direct and indirect, in which case mobilization must be added to the various category totals essentially as a bid item. But since owners might object to paying for mobilization, these costs then must be distributed, usually to the bid items that benefit from the mobilization.

4. Apportion the indirect expenses. In Figure 6.26 these are apportioned based on estimated direct labor expenses, and in Figure 6.27 they are apportioned based on estimated total direct expenses. For example, in Figure 6.26 the indirect expenses apportioned to excavation are calculated using the formula $73,034 (total indirect expenses) divided by $107,413 (total direct labor expenses) times $1404 (the direct labor expenses) for this item. This comes to $955. For the apportionment based on total direct expenses (Fig. 6.27), the calculation would be $73,034 divided by $292,137 times $9316. This comes to $2329, a significant difference. Using the first method, tasks that are heavy in labor (e.g., rough carpentry) are apportioned greater amounts of indirect expenses.

5. Add the direct and indirect portions of the cost of each task to obtain the total cost of that task and enter these in the "total" column. The sum of this column is your total estimated cost ($292,137 + 73,034 = $365,172). You should cross-check this against your original calculations of direct and indirect costs.

6. Divide the total cost of each task by the units of quantity on which the bid is to be based and enter in the "total unit price" column. If the owner is requesting unit prices, these are the unit costs on which the unit bid prices will be based.

The summary form provides you with four more columns to allow you to adjust your bid. Here, based on your assessment of your competition and client, you may add to or subtract from your estimate. Some contractors choose to play with the numbers and unbalance their bids; I recommend against this practice.

6.12 THE 27 TIMES TABLE

There are 27 cu. ft. in one cu. yd. In elementary school you learned the 2 through 12 times tables. As an estimator you may find it convenient to also learn the 27 times table to help you convert from cu. ft. to cu. yd. and vice versa.

$$1 \times 27 = 27$$
$$2 \times 27 = 54$$
$$3 \times 27 = 81$$
$$4 \times 27 = 108$$
$$5 \times 27 = 135$$
$$6 \times 27 = 162$$
$$7 \times 27 = 189$$
$$8 \times 27 = 216$$
$$9 \times 27 = 242$$
$$10 \times 27 = 270$$
$$11 \times 27 = 297$$
$$12 \times 27 = 324$$

6.13 BOARD MEASURE (BM)

Multiply linear ft. (LF) × constant to obtain board measure (BM).

Size	Constant	Size	Constant
1 × 3	¼	3 × 12	3
1 × 4	1/3	4 × 4	1 1/3 or 4/3
1 × 6	½	4 × 6	2
1 × 8	2/3	4 × 8	2 1/3 or 8/3
1 × 10	5/6	4 × 10	3 1/3 or 10/3
1 × 12	1	4 × 12	4
2 × 3	½	8 × 8	5 1/3 or 16/3
2 × 4	2/3	10 × 10	8 1/3 or 25/3
2 × 8	1 1/3 or 4/3	12 × 12	12
2 × 10	1 2/3 or 5/3	14 × 14	16 1/3 or 49/3
2 × 12	2	16 × 16	21 1/3 or 64/3
3 × 3	¾	18 × 18	27
3 × 4	1	20 × 20	33 1/3 or 100/3
3 × 6	1½ or 3/2	22 × 22	40 1/3 or 121/3
3 × 8	2	24 × 24	48
3 × 10	2½ or 5/2		

6.14 INCHES TO DECIMALS OF 1 FOOT

Some people like to convert inches to decimals of 1 foot. This makes multiplication easier. On the down side you must make more conversions and multiplications, thereby raising the risk of error.

Inches to Decimals of 1 Foot			
1.0 =	0.083	6.5 =	0.5417
1.5 =	0.125	7.0 =	0.583
2.0 =	0.1667	7.5 =	0.625
2.5=	0.2087	8.0 =	0.667
3.0 =	0.25	8.5 =	0.708
3.5 =	0.2917	9.0 =	0.75
4.0 =	0.333	9.5 =	0.792
4.5 =	0.375	10.0 =	0.833
5.0 =	0.417	10.5 =	0.875
5.5 =	0.458	11.0 =	0.917
6.0 =	0.5	11.5 =	0.958

6.15 CONVERSION FACTORS

Multiply	by	to obtain
acres	43,580	square feet
acres	4047	square meters
acres	1562×10^3	square miles
acres	4840	square varas[3]
acre-feet	43,560	cubic feet
acres	100	square meters
board feet	144 sq. in. × 1 in.	cubic inches
British Thermal Units (Btus)	0.2520	kilogram-calories
Btus	777.5	foot-pounds
Btus	2.928×10^4	kilowatt-hours
Btus per minute	0.02358	horsepower
Btus per minute	0.01757	kilowatts
Btus per minute	17.57	watts
centares	1	square meters
centigrams	0.01	grams
centiliters	0.01	liters
centimeters	0.3937	inches

3. The vara is a Texas unit of length equal to 33.33 inches (84.66 centimeters).

centimeters	0.01	meters
centimeter-grams	10	millimeters
cord-feet	4 ft. × 4 ft. × 1 in.	cubic feet
cords	8 ft. × 4 ft. × 4 ft.	cubic feet
cubic centimeters	6.102×10^{-2}	cubic inches
cubic centimeters	10^{-6}	cubic meters
cubic centimeters	2.642×10^{-4}	gallons
cubic centimeters	10^{-3}	liters
cubic feet	2.832×10^{4}	cubic centimeters
cubic feet	1728	cubic inches
cubic feet	0.02832	cubic meters
cubic feet	0.03704	cubic yards
cubic feet	7.481	gallons
cubic feet	28.32	liters
cubic feet per minute	475.0	cubic cm per second
cubic feet per minute	0.1247	gallons per second
cubic feet per minute	0.4720	liters per second
cubic feet per minute	62.4	pounds water per minute
cubic inches	16.39	cubic centimeter
cubic meters	10^{6}	cubic centimeter
cubic meters	35.31	cubic feet
cubic meters	1.308	cubic yards
cubic meters	264.2	gallons
cubic yards	27	cubic feet
cubic yards	0.7646	cubic meters
cubic yards	202.0	gallons
cubic yards per minute	0.45	cubic feet per second
cubic yards per minute	3.367	gallons per second
degrees (angle)	60	minutes
degrees (angle)	0.01745	radians
degrees (angle)	3600	seconds
ergs	9.486×10^{-11}	Btus
fathoms	6	feet
feet	0.3048	meters
feet	0.36	varas
feet	1/3	yards
feet of water	0.4335	pounds per square inch
feet per minute (fpm)	0.5080	centimeters per second
fpm	0.01136	miles per hour

feet per second (fps)	1.097	kilometers per hour
fps	0.5921	knots per hour
fps	18.29	meters per minute
feet per 100 ft.	1	percent grade
foot-pounds	1.286×10^{-3}	Btus
foot-pounds	1.356×10^{7}	ergs
foot-pounds	5.050×10^{-7}	horsepower hours
foot-pounds per minute	1.286×10^{-3}	Btus per minute
foot-pounds per minute	2.260×10^{-5}	kilowatts
furlong	40	rods
gallons	3785	cubic centimeters
gallons	0.1337	cubic feet
gallons	231	cubic inch
gallons	3.785×10^{-3}	cubic meters
gallons	4.951×10^{-3}	cubic yards
gallons per minute	2.228×10^{-3}	cubic feet per second
grams	980.7	dynes
grams	10^{-3}	kilograms
grams	10^{3}	milligrams
grams	0.03527	ounces
grams	0.03215	ounces (troy)
grams	2.205×10^{-3}	pounds
gram-calories	3.968×10^{-3}	Btus
grams per centimeter	5.600×10^{-3}	pounds per inch
grams per cubic centimeter	62.43	pounds per cubic feet
hectares	2.471	acres
horsepower	42.44	Btus per minute
horsepower	33,000	foot-pounds per minute
horsepower	550	foot-pounds per second
horsepower	1.014	horsepower (metric)
horsepower	0.7457	kilowatts
horsepower	745.7	watts
inches	2.540	centimeters
inches	0.03	varas
inches of water	0.5781	ounces per square inch
inches of water	5.204	pounds per square foot
inches of water	0.03613	pounds per square inch
joules	9.486×10^{-4}	Btus
joules	10^{7}	ergs

joules	0.7376	foot-pounds
joules	2.778×10^{-4}	watt-hours
kilograms (kg)	980,665	dynes
kg	10^3	grams
kg	2.2046	pounds
kg	1.102×10^{-3}	tons (short)
kg-calories	3.968	BTU
kg-calories	3.968	foot-pounds
kg-calories	1.162×10^{-3}	kilowatt-hours
kg-meters	9.302×10^{-3}	Btus
kg-meters	9.807×10^7	ergs
kg per cubic meter	10^{-3}	grams per cubic centimeter
kg per cubic meter	0.06243	pounds per cubic foot
kg per square meter	0.2048	pounds per square foot
kg per square meter	1.422×10^{-3}	pounds per square inch
kilometers (km)	3281	feet
km	0.6214	miles
km per hour	0.5396	knots per hour
kilowatts	56.92	Btus per minute
kilowatts	4.425×10^4	foot-pounds per minute
kilowatts	1.341	horsepower
kilowatt-hours	3415	Btus
kilowatt-hours	2.655×10^6	foot-pounds
knots	1.853	km per hour
knots	1.152	miles per hour
links (engineer's)	12	inches
links (surveyor's)	7.92	inches
liters	10^3	cubic centimeters
liters	0.2642	gallons
liters	1.057	quarts (liquid)
liters per minute	5.885×10^{-4}	cubic feet per second
liters per minute	4.403×10^{-3}	gallons per second
meters	100	centimeters
meters	3.2808	feet
meters	39.37	inches
meters	10^{-3}	km
meters	10^3	millimeters
meters	1.0936	yards
miles	5280	feet

miles	1.6093	kilometers
miles	1760	yards
miles per hour	1.467	feet per second
miles per hour	1.6093	kilometers per hour
miles per hour	0.8684	knots per hour
millimeters	0.1	centimeters
millimeters	0.03937	inches
mils	10^{3}	inches
minutes (angle)	2.909×10^{4}	radians
minutes (angle)	60	seconds (angle)
nautical miles	1.152	miles
nautical miles	2207	yards
ounces	8	drams
ounces	437.5	grains
ounces	28.35	grams
ounces	0.0625	pounds
ounces (fluid)	1.805	cubic inches
perches (masonry)	24.75	cubic feet
pints (dry)	33.60	cubic inches
pints (liquid)	28.87	cubic inches
pounds	444,823	dynes
pounds	453.6	grams
pounds	16	ounces
pound-feet	1.356×10^{7}	centimeter-dynes
pound-feet	13,825	centimeter-grams
pound-feet	0.1383	meters-kilogram
pounds of water	0.01602	cubic feet
pounds of water	27.68	cubic inches
pounds of water	0.1198	gallons
pounds per cubic foot	16.02	kilogram per cubic meter
pounds per cubic foot	27.68	grams per cubic centimeter
pounds per foot	1.488	kilogram per meter
pounds per square foot.	0.01602	feet of water
pounds per square inch	2.036	inches of mercury
pounds per square foot	4.882	kilogram per square meter
pounds per square inch	2.307	feet of water
pounds per square inch	703.1	kilogram per square meter
pounds per square inch	144	pounds per square foot
quadrants (angle)	90	degrees

quadrants (angle)	5400	minutes
quadrants (angle)	1.571	radians
quarts (dry)	67.20	cubic inches
quarts (liquid)	57.75	cubic inches
radians	57.30	degrees
radians	3438	minutes
radians	0637	quadrants
reams	500	sheets
revolutions	360	degrees
revolutions per minute (rpm)	6	degrees per second
rpm	0.1047	radians per second
rpm	0.0167	revolutions per second
revolutions per second	360	degrees per second
revolutions per second	6.283	radians per second
rods	16.5	feet
seconds (angle)	4.848×10^{-6}	radians
square centimeters	0.1550	square inches
square centimeters	100	square millimeters
square feet	2.296×10^{-5}	acres
square feet	0.09290	square meters
square feet	$3.587 \, 10^{-8}$	square miles
square feet	0.1296	square varas
square feet	1/9	square yards
square feet	6.452	square centimeters
square inches	6.944×10^{-3}	square feet
square kilometers	247.1	acres
square kilometers	10.76×10^{6}	square feet
square kilometers	10^{6}	square meters
square kilometers	0.3861	square miles
square kilometers	1.196×10^{6}	square yards
square kilometers	2.471×10^{-4}	acres
square meters	10.764	square feet
square meters	3.861×10^{-7}	square miles
square meters	1.196	square yards
square miles	640	acres
square miles	27.88×10^{6}	square feet
square miles	2.590	square kilometers
square miles	3.098×10^{6}	square yards
square yards	2.066×10^{-4}	acres

square yards	9	square feet
square. yards	0.8361	square meters
square yards	3.228×10^{-7}	square miles
temp. (°C) +273	1	absolute temp. (°C)
temp. (°C) + 17.8	1.8	temp. (°F)
temp. (°F) + 460	1	absolute temp. (°F)
temp. (°F) − 32	5/9	temp. (°C)
tons (long)	1016	kilograms
tons (long)	2240	pounds
tons (metric)	10^3	kilograms
toms (metric)	2205	pounds
tons (short)	907.2	kilograms
tons (short)	2000	pounds
tons (short) per square foot	9765	kilograms per square meter
tons (short) per square foot	13.89	pounds per square inch
tons (short) per square foot	1.406×10^6	kilograms per square meter
tons (short) per square foot	2000	pounds per square inch
watts	0.05692	Btus per minute
watts	10^7	ergs per second
watts	44.26	foot-pounds per minute
watts	1.341×10^{-3}	horsepower
watts	10^2	kilowatts
watt-hours	3.415	Btus
weeks	168	hours
yards	91.44	centimeters
yards	3	feet
yards	36	inches
yards	0.9144	meters

6.16 SIZING ELECTRICAL WIRING

Most project plans include an electrical design.[4] The plans may, however, assume that a licensed electrician will properly select

---

4. The information in this section comes primarily from Timothy Thiele, *What Size Electrical Wire Do I Need* and *Determining Proper Electrical Wire Size For Underground Circuit Cable Length*. These can be found on the web at http://electrical.about.com/od/wiringcircuitry/a/electwiresizes.htm and http://electrical.about.com/od/wiringcircuitry/qt/wiresizeandcablelength.htm.

the correct wires for certain applications. This, of course, does not help you estimate the cost of wiring. Lots can go wrong with electric wiring; therefore don't allow yourself to get in over your head. However, if you must complete your estimate and have no one to turn to, here are some guidelines for selecting the wire sizes for pricing. Again, don't install wiring unless you are certain it has been sized properly.

Your choice of conductor will depend on what the wire will feed. Table 6.1 provides the minimum wire gauges for various family home uses. Wire ampacity[5] is the current, in amperes, that a conductor can carry continuously under conditions of use without exceeding its temperature rating. The selected ampacity should be compatible with the circuit breaker or fuse that is protecting the conductor.

When considering the appropriate wire size for a circuit, you need to consider circuit length. All electrical wire has resistance, and as the cable length increases so too does resistance, causing the voltage to drop. Resistance per ft. decreases as the diameter of the wire increases. (Think of a water pipe. As the diameter of the pipe increases, its cross section increases by radius squared, thereby letting far more water to pass through it. Similarly, as wire diameter increases its area also increases by radius squared, thereby letting more current through it.) Therefore if in doubt you should choose the next-larger-sized ampacity wire size to lessen the effects of wire resistance. For example, on a 20-amp circuit, you would normally use 12-gauge wire to supply the circuit. However, if the run is over 50

---

5. Ampacity as defined in Article 100 of the Definitions of the 2002 National Electric Code is the maximum current, in amperes, that a conductor can carry continuously under conditions of use without exceeding its temperature rating. Wire diameter increases as the wire gauge number decreases.

Table 6.1. Wire gauges and uses		
Wire use	**Rated ampacity**	**Wire gauge**
Low-voltage lighting and lamp cords	10 amps	18 gauge
Extension cords	13 amps	16 gauge
Light fixtures, lamps, lighting runs	15 amps	14 gauge
Receptacles, 110-volt air conditioners, sump pumps, kitchen appliances	20 amps	12 gauge
Electric clothes dryers, 220-volt window air conditioners, built-in ovens, electric water heaters	30 amps	10 gauge
Cooktops	45 amps	8 gauge
Electric furnaces, large electric heaters	60 amps	6 gauge
Electric furnaces, large electric water heaters, subpanels	80 amps	4 gauge
Service panels, subpanels	100 amps	2 gauge
Service entrance	150 amps	1/0 gauge
Service entrance	200 amps	2/0 gauge

ft. long, you could use a 10-gauge wire instead, which is rated for a 30-amp circuit.[6]

American Wire Gauge (AWG) sizes may be determined by measuring the diameter of the conductor (the bare wire) with the insulation removed. Refer to Table 6.3 for dimensions. When choosing wire gauge, the distance the wire must run and the amperage it will be expected to carry must be determined first. Note that you can always use thicker wire (a lower gauge number) than is recommended. The wire

---

6. You should also refer to your locally enacted electrical code regarding allowable voltage drops.

gauge table (Table 6.3) can be found at http://www.rbe electronics.com/wtable.htm.

Line loss is the voltage drop between the electrical service and the load. Line loss usually controls wire size determination rather than the allowable ampacities listed in Chapter 3 of the National Electric Code. Resistance and its inevitable voltage drop causes wires to get hot, and the lower voltage causes the equipment to which they are attached to run less effectively. Therefore it is important that you not undersize wiring.

Now even more of the constructions materials that we purchase are being imported. This includes electrical wires. Table 6.2 provides you with a conversion table from metrically sized wires to American Wire Gauge (AWG) with which you are probably more familiar. Table 6.3 converts AWG sizes to wire diameters in inches.

Table 6.2. Metric to AWG conversion table	
Metric size (mm^2)	AWG size
0.5	20
0.8	18
1.0	16
2.0	14
3.0	12
5.0	10
8.0	8
13.0	6
19.0	4
32.0	2
52.0	0

Table 6.3. Wire gauge diameter table	
AWG	**Wire diameter in inches**
20	0.03196118
18	0.040303
16	0.0508214
14	0.064084
12	0.08080810
10	0.10189
8	0.128496
6	0.16202
5	0.18194
4	0.20431
3	0.22942
2	0.25763
1	0.2893
0	0.32486
00	0.3648

There are websites on which you can find additional information about electrical installations. A good place to start is Timothy Thiele's *What Size Electrical Wire Do I Need* and *Determining Proper Electrical Wire Size For Underground Circuit Cable Length*. These can be found on the web at http://electrical.about.com.

6.17 MISCELLANEOUS ROUGH CARPENTRY

Framing lumber is always measured in board measure (BM), a system of units for measuring lumber based on the board ft. Unless otherwise stated in the takeoff, the framing will be dressed on four sides (D4S). Note that rough framing materials

may be less costly to purchase, but they are more costly to handle than dressed stock. Always check the specifications. If the
roof boarding is to be 1-in. finish thickness, then 1.25 board is
called for; if the roof boarding is to be 1-in. nominal thickness,
you can use 1-in. stock.

Boarding for floors, roofs, and wall sheathing require that
you allow for cutting and waste depending on the type of board,
its width, and whether it is to be laid straight or diagonally. The
amount of waste involved in cutting and fitting varies but is usually in the range of 7.5% to 12.5%. Tongue and groove (T&G)
lays ¾ in. less than its nominal. Therefore, for 1 × 6s, a loss of
¾ in. in 6 in. amounts to almost 14%. Square-edge boarding
would lay 3/8 in. to ½ in. less than its nominal width.

When cutting hips and roof rafters, allowance must be made
for both the eaves overhang and the long bottom cut.

Laminated wood beams, purloins, and so forth are made
and shipped, as are arches. Your takeoff should be by piece by
size. Usually the subbid for supply and delivery will include the
necessary bolts, angles, plates, and so on, but this should not be
presumed. The setting of laminated wood framing will probably
require a crane. Also include cost of railroad siding, trucking to
the job, and unloading if not included in the subbid.

Stud framing is usually reduced to BM, although you may
find it better to leave miscellaneous minor items of studding in
linear ft. (LF) when the stock is not larger than 2 × 3. LF will
warn you that a great amount of cutting may be required.

All boarding is usually measured in sq. ft. (SF). Waste must
be carefully considered.

6.18 MISCELLANEOUS FINISH CARPENTRY

Moldings are taken off in linear ft., with each member measured separately. Trim that is curved or polygonal in shape should be separate in the takeoff from ordinary moldings. Paneling is measured in sq. ft., separating paneling up to 4 ft., 4 to 6 ft., and over 6 ft. The low dado paneling will cost more to set per sq. ft. than similar paneling 6 ft. high. Paneling to be built up on the wall is measured as separate items: plywood, cover strips, and moldings. Prefabricated paneling, on the other hand, is measured in sq. ft. of finished panel, with only the base and caps molding separately measured.

Cabinets and casework are usually taken off in items of stated size. Casework that cannot be preassembled should be kept separate from casework that will be completely prefabricated.

Cutouts in countertops should be taken off as labor items unless you are certain the mill will make the cutouts.

6.19 BOARD MEASURE, SIZE, AND WEIGHT

Table 6.4 provides information about commercial lumber sizes available in the United States. Lumber quantities are expressed in ft., board measure (BM or ft. b.m.), in board ft. (bd. ft.), or in thousand board ft. (M bd. ft.). One board ft. is the amount of lumber in a rough-sawed board 1 ft. long, 1 ft. wide, and 1-in. thick (144 cu. in.) or the equivalent volume in any other shape (e.g., 1 ft. long × 6 in. wide × 2 in. thick). The originals, or "nominal" dimensions, and volumes determine the number of board ft. in a quantity of dressed lumber. The process of surfacing or other machining does not lessen the board ft. that are sold. Under American standards a board designated as a 1-in. × 12-in. board is in fact 25/32 in. × 11½ in. Therefore one hundred 1 in. dressed boards 16 ft. long contain

(100 boards) × (1 in. thick × 12 in. wide)
(16 ft. long × 12 in. per ft.) = cu. in. ÷ 144 = BM,

or in this case,

$$(100) \times (1 \times 12) \times (16 \times 12) \div 144 = 1600 \text{ board ft.},$$

but have an actual volume of only

$$(100) \times (25/32 \times 11.5) \times (16 \times 12) \div 144 = 1198 \text{ cu.ft.}$$

If you are thinking of area coverage,

$$100 \times 1 \text{ in.} \times 16 \text{ in.} = 1600 \text{ sq. ft.}$$

versus

$$100 \times (11.5 \div 12) \times 16 = 1533 \text{ sq. ft.}$$

Weights were calculated assuming the weight of wood to be 35 pounds per cu.ft. For example, an 8 × 12 has an area section of 86.25 sq. in. Multiply this by 12 in. (1 ft.) to get 1035 cu. in. Divide 1035 by 1728 (the number of cu. in. in a cu. ft.) to get 0.599 cu. ft. Multiply 0.599 by 35 to get 20.964.

Note that the density (pounds per cu. ft.) of wood varies with species. For example, poplar has a density of about 30 pounds/cu. ft., white pine is 26 pounds/cu. ft., and Ponderosa pine is 28 pounds/cu. ft. Weight also varies with moisture content. In some species, nominal sizes may vary from the chart (e.g., a 2 × 4 may be 1 5/8 × 3 5/8). You can find additional information regarding the weight of construction materials online and in estimating guides such as Dewalt's *Construction*

Table 6.4. Commercial lumber sizes (for lumber commercially available in the United States)			
Nominal size	Actual size	Area of section in sq. in.	Weight per ft. (lbs.)
1 × 3	0.75 × 2.5	1.875	0.46
1 × 4	0.75 × 3.5	2.625	0.64
1 × 6	0.75 × 5.5	4.125	1.00
1 × 8	0.75 × 7.25	5.438	1.32
1 × 10	0.75 × 9.25	6.938	1.69
1 × 12	0.75 × 11.25	8.438	2.05
2 × 3	1.5 × 2.5	3.750	0.94
2 × 4	1.5 × 3.5	5.250	1.28
2 × 6	1.5 × 5.5	8.250	2.01
2 × 8	1.5 × 7.25	10.875	2.64
2 × 10	1.5 × 9.25	13.875	3.37
2 × 12	1.5 × 11.25	16.875	4.10
2 × 14	1.5 × 13.25	19.875	4.83
3 × 3	2.5 × 2.5	6.250	1.52
3 × 4	2.5 × 3.5	8.750	2.13
3 × 6	2.5 × 5.5	13.750	3.34
3 × 8	2.5 × 7.25	18.125	4.41
3 × 10	2.5 × 9.25	23.125	5.62
3 × 12	2.5 × 11.25	28.125	6.84
3 × 14	2.5 × 13.25	33.125	8.05
3 × 16	2.5 × 15.25	38.125	9.27
4 × 4	3.5 × 3.5	12.250	2.98
4 × 6	3.5 × 5.5	19.250	4.68
4 × 8	3.5 × 7.25	25.375	6.17
4 × 10	3.5 × 9.25	32.375	7.87

Table 6.4. Commercial lumber sizes
(for lumber commercially available in the United States) (*continued*)

Nominal size	Actual size	Area of section in sq. in.	Weight per ft. (lbs.)
4 × 12	3.5 × 11.25	39.375	9.57
4 × 14	3.5 × 13.25	46.375	11.27
4 × 16	3.5 × 15.25	53.375	12.97
6 × 6	5.5 × 5.5	30.250	7.35
6 × 8	5.5 × 7.5	141.250	10.03
6 × 10	5.5 × 9.5	52.250	12.70
6 × 12	5.5 × 11.5	63.250	15.37
6 × 14	5.5 × 13.5	74.250	18.05
6 × 16	5.5 × 15.5	85.250	20.72
6 × 18	5.5 × 17.5	96.250	23.39
8 × 8	7.5 × 7.5	56.250	13.67
8 × 10	7.5 × 9.5	71.250	17.32
8 × 12	7.5 × 11.5	86.250	20.96
8 × 14	7.5 × 13.5	101.250	24.61
8 × 16	7.5 × 15.5	116.250	28.26
8 × 18	7.5 × 17.5	131.250	31.90
10 × 10	9.5 × 9.5	90.250	21.94
10 × 12	9.5 × 11.5	109.250	26.55
10 × 14	9.5 × 13.5	128.250	31.17
10 × 16	9.5 × 15.5	147.250	35.79
10 × 18	9.5 × 17.5	166.250	40.41
12 × 12	11.5 × 11.5	132.250	32.14
12 × 14	11.5 × 13.5	155.250	37.73
12 × 16	11.5 × 15.5	178.250	43.32

Estimating, Professional Reference by Adam Ding and *Walker's Building Estimating Reference Book.*[7]

6.20 NAILS AND FASTENERS

Table 2304.9.1 of the International Building Code, 2006, provides fastening schedule. Other localities may provide their specific requirements that you will have to follow. You may find the following two figures handy to have close at hand.

Online are numerous tables and calculators to help you determine the type and amount of nails to purchase.[8] Here, Figure 6.28 provides commercial nail sizes and Figure 6.29 provides commercial nail uses.

Figure 6.28. Commercial nail sizes

SIZE	LENGTH IN	COMMON GAGE	DIAMETER (D) NO./LB	DIAMETER (D) INCHES	FINISHING GAGE	FINISHING NO./LB	FLOORING GAGE	FLOORING NO./LB
2d	1	15	876	0.072	16.5	1,351		
3d	1.25	14	568	0.083	15.5	807		
4d	1.5	12.5	316	0.109	15	584		
5d	1.75	12.5	271	0.109	15	500		
6d	2	11.5	181	0.113	13.5	309	11	157
8d	2.5	10.3	106	0.131	12.5	189	10	99
10d	3	9	69	0.148	11.5	121	9	69
12d	3.25	9	64	0.155	11.5	113	8	54
16d	3.5	8	49	0.162	11	90	7	43
20d	4	6	31	0.203	10	62	6	31
30d	4.5	5	24	0.207				
40d	5	4	18	0.225				
60d	6	2	11	0.263				
SPIKES								
7"	7"	5/16"	7	5/16"				
8"	8"	3/8"	6	3/8				
9"	9"	3/8"	5	3/8"				
10"	10"	3/8"	4	3/8"				
12"	12"	3/8"	3	3/8"				

7. This book is intended to provide you with a process to follow as you prepare your estimate. For information on actual pricing there are numerous references available. These can be found on line by going to Amazon, Barnes and Nobel, and vendors of estimating software.

8. For examples, see http://www.onlineconversion.com/common_nails_smooth _shank.htm and http://www.plm.com/nails.htm.

Figure 6.29. Commercial nail uses

STANDARD NAILS

Rough Carpentry	Penny	Inches	Type of Nail
1" stock	8	2.5	Common nail
2"	16 - 20	3.5 or 4	Common nail
3"	40 - 60	5 or 6	Common nail or spike
Concrete forms	varies		Double headed or common nail
General framing & large members	10 - 60	3 -6	Common nail or spike
Toe nailing studs, joists, etc.	10	3	Common nail
Spiking plates and sills	16	3.5	Common nail
Toe nailing rafters andplates	10	3	Common nail, may be zinc coated
Sheathing roof and walls	10	3	Common nail, may be zinc coated

Finish Carpentry			
Moldings	As required		Brads
Carpet strips, shoes	8	2.5	Common nail, may be zinc coated
Door and window stops, 1/4" - 1/2"	4	1.5	Finishing or casing nails
Ceiling, trim, base balusters, 1/2" -3/4	6	2	Finishing or casing nails
Door and window trim, 1" - 1-1/4"	8	2.5	Finishing or casing nails
Door siding, 1"	10	3	Finishing or casing nails
Drop siding, 1"	7 or 9	2.25 -2.75	Siding nail (7d), Casing nail (9d)
Bevel siding, 1/2"	6 or 8	2 or 2.5	Finishing nail (6d) or Siding nail (8d)

Lathing			
Wood lath	3	1.25	Blue lath nail
Gypsum lath	3	1.25	Blue common nail
Metal lath, interior		1	Blue lath nail, staples, offset headed nail
Metal lath, exterior	3	1.25	Self furring nail, staples, cement coated

Sheathing or Siding			
Fiber board 1/2"		1.5 - 2	Galvanized roofing w/7/16 head
Gypsum board 1/2"		1.75	Galvanized roofing w/7/16 head
Plywood 5/15" and 3/8"	6	2	Common nail
Plywood 1/2" and 5/*"	8	2.5	Common nail

Roofing and Sheet Metal			
Aluminum roofing	1	1.75 - 2.5	Aluminum nail w/neoprene washer
Asphalt shingles		1 - 1.25	Galvanized large head roofing nail
Copper cleats & flashing to wood			Copper wire or cut slating nail
Clay tile	4 - 6	1.5 - 2	Barbed copper nail
Prepared felt roofing		1 - 1.25	Zinc roofing, large head roofing
Slate			Copper wire slating, large head
			In dry climates, zinc or copper clad nail
Tin, zinc roofing			Zinc coated nail (roofing or slating)
monel roofing			Monel nail

7. GLOSSARY[1]

This glossary is extensive. Though some of the words may seem basic, I did not intend to talk down to you. Rather, it is my intension to provide you with an additional tool to use in your role as an estimator.

Most of the words herein relate to building construction and renovation work as opposed to heavy construction. There are also some banking, contract, and specification terms that you may find helpful. And with ever-increasing emphasis being placed on "green" buildings and "sustainable" building products, words relating to these terms have also been included.

You should also be aware that some construction terms vary from place to place. If you notice a variation in definition, pencil in your local meaning.

2/10 Rule The rule states that the top of a flue must be at least 2 ft. higher than any roof part within 10 ft.

4/1 Rule A rule for safe placement of a ladder. The 4/1 Rule states that for every 4 ft. of working ladder length, the base of the ladder should be 1 ft. out from the top support point.

---

1. This glossary was assembled using these references: Home Building Manual, "Construction Glossary from Home Building Manual," http://www.homebuilding manual.com/Glossary.htm; Contractorslicense.com, "Construction Glossary and Terms," http://www.contractorslicense.com/0-24-glossary.htm; US Green Building Council, *Existing Buildings: Operations & Maintenance Reference Guide*, 1st ed., August 2008; International Code Council, *International Residential Code For One- and Two-Family Dwellings, 2006* (Falls Church, VA: International Code Council, Inc., 2006); *New York City Electrical Code 2004* based on 2002 National Electric Code, International Electrical Code Series; and common sense.

80/20 Rule (Pareto's Rule) The 80/20 Rule means that in anything a few (20%) are vital and many (80%) are trivial. In Pareto's case, it meant 20% of the people owned 80% of the wealth. In Juran's initial work, he identified 20% of the defects causing 80% of the problems. Project managers know that 20% of the work (the first 10% and the last 10%) consumes 80% of your time and resources. You can apply the 80/20 Rule to almost anything, from the science of management to the physical world.

AC An abbreviation for alternating current; the type of electrical current provided to most homes.

AC or A/C An abbreviation for air-conditioner or air-conditioning.

A/C condenser The outside fan unit of an air-conditioning system. It removes the heat from the Freon gas, "turns" the gas back into a liquid, and pumps the liquid back to the coil in the furnace.

A/C disconnect The main electrical on-off switch near the A/C condenser.

Acceleration A situation that forces a contractor to increase the work effort to meet the contract completion date and avoid possible liquated damages.

Access A means of approach to a structure or a part thereof such as a road, street, pathway, or corridor.

Access floor A floor structure normally constructed over the floor slab that allows access for cabling and ducts; also referred to as raised flooring.

Accessibility Generally refers to a facility's capability to permit disabled people to enter and use the room or building. Chapter 11 of the International Building Code is dedicated to accessibility.

Accessible (as applied to equipment) Capable of being approached closely and not guarded by locked doors, elevation, or other effective means.

Accessible (as applied to wiring methods) Capable of being removed or exposed without damaging the building structure or finish, or not permanently closed in by the structure or finish of the building.

Accessible, readily (readily accessible) Capable of being reached quickly for operation, renewal, or inspections without requiring those to whom ready access is requisite to climb over or remove obstacles, or to resort to portable ladders, etc.

Accident frequency The ratio of accidents and hours worked. While the ratio in most industries is usually calculated per million hours worked, smaller firms can use fewer hours worked to make comparisons between projects and project managers.

Acetal plastic A type of plastic used in polybutylene (PB) pipe.

Addition An extension or increase in floor area or height of a building that increases its exterior dimensions.

Administrative approval The processing and proper approval of a work request, change order, and the like to ensure that funding and level of effort increases (decreases) will be met.

Administrative services Generally, services that are not directly construction related (e.g., transportation, food services, security, office equipment, etc.).

Admixtures Material—other than water, cement, aggregate, or fiber reinforcement—used as an ingredient in a batch of concrete or mortar. Among the most common admixtures are those that improve plasticity, retard or advance hydration, and add color.

A-E or A/E An architect and engineering firm; typically used to identify the principal designer(s) on a project.

Aerator The round, screened screw-on tip of a sink spout. It mixes water and air for a smooth flow.

Aggregate Sand, gravel, or a combination of both. The use and size of gravel varies depending on whether the product is concrete (sand plus gravel larger than ¼ in.), grout (sand plus gravel ¼ in. or smaller), mortar or plaster (sand only).

Air break A piping arrangement in which a drain from a fixture discharges through the atmosphere into a receptacle. This arrangement prevents the creation of a siphon.

Air-conditioning The process that simultaneously controls temperature, humidity, movement, cleanliness, and odor of air circulated through a space.

Air-conditioning coil A coil configured to remove heat from refrigerant gas.

Air-cooled flue A flue that may consist of double or triple pipes in which heated gases rising out of the inner flue create a draft that pulls cool air through the outer pipe(s).

Air-entrainment agent A type of admixture. This is actually a detergent that produces small, evenly spaced bubbles in concrete. It makes the concrete more plastic or workable and more frost resistant.

Air gap (drainage system) The unobstructed vertical distance through air between a wastewater pipe outlet and the flood-level rim of the receptor into which it is discharging. An air gap prevents wastewater from backing up into the originating wastewater pipe.

Air gap (water-distribution system) The unobstructed vertical distance through air between the lowest opening from a water supply discharge to the flood-level rim of a plumbing fixture.

Air space The area between insulation facing and interior of exterior wall coverings. This is normally a 1-in. air gap.

Albedo Synonymous with solar reflectance.

Allocate To set aside funds for a project.

Allowable soil pressure The maximum stress permitted in soil of a given type under given conditions.

Allowable stress The maximum stress permitted at a given point in a structural member under given conditions.

Allowance(s) A sum of money set aside in the construction contract for items that have not been selected and specified in the construction contract. For example, selection of tile for flooring may require an allowance for an underlayment material or an electrical allowance that sets aside an amount of money to be spent on electrical fixtures.

All-purpose mud A type of drywall mud containing adhesive chemicals that hold the drywall tape in place and help the mud adhere to the drywall face.

Alterations Any construction or renovations to an existing structure other than repair or addition. Also, the moving of a building is frequently deemed an alteration.

Alternate-fuel vehicles Vehicles that use low-polluting, nongasoline fuels such as electricity, hydrogen, propane, or compressed natural gas, methanol, and ethanol. Efficient gas-electric hybrid vehicles are included in that group.

Aluminum oxide A chemical that is created when aluminum is exposed to the atmosphere. Aluminum oxide is an insulator, and when it builds up on an aluminum power wire, it can create heat that can loosen the connection, cause arcing and even fire.

Ambient lighting The light reaching an object in a room from all light sources; surrounding light.

American standard gauge Units used to measure the size of wire. The smaller the wire, the larger its gauge number will be. For example, an 18-gauge wire is a very thin wire used for a doorbell, and a 2 gauge is a large feeder wire. Wires larger than 2 gauge are measured in aughts. Wire larger than four aught are measured in microcircular-mills (MCM).

American wallpaper roll A roll of wallpaper that contains about 36 sq. ft.

Amortization A payment plan by which a loan is reduced through monthly payments of principal and interest.

Ampacity The maximum current, in amperes, that a conductor can carry continuously under the conditions of use without exceeding its temperature rating.

Anchor bolt A bolt used to connect the foundation to the inside of the framing above it. The anchor bolts are placed in the foundation while the concrete is still wet; then, after the concrete cures, the foundation sill plate is attached to the anchor bolts.

Angle iron L-shaped metal piece commonly used as a lintel.

Annual percentage rate (APR) The annual cost of credit over the life of a loan, including interest, service charges, points, loan fees, mortgage insurance, and other items.

Antioxidant compound A compound applied onto aluminum wires to prevent aluminum oxide from forming.

Antistratification device A device that stirs water in a hot water heater

to prevent the stratification of hotter water at the top and cooler water at the bottom.

APP (Atactic polypropylene) Plasticizer used in the torch-down type of modified bitumen roof systems.

Appliance Utilization equipment, generally other than industrial, that is normally built in standardized sizes or types and is installed or connected as a unit to perform one or more functions such as clothes washing, air conditioning, food mixing, deep frying, etc.

Appraisal An expert valuation of property.

Approved Acceptable to the authority having jurisdiction.

Approved agency An established and recognized agency that is regularly engaged in conducting tests or furnishing inspection services and has been approved by the Department of Buildings and Safety or similar organization.

Approved vendor list A current list of vendors providing the owner with goods and services. Many government agencies award contracts only to firms on their approved vendor list.

Apron A trim board that is installed beneath a window sill.

Arbitration The settlement of a contract dispute by selecting an impartial third party to hear both sides and reach a decision. Many contracts include a clause that requires all disputes be resolved through arbitration. Before signing such a contract understand that by so doing you are giving up your right to sue in court.

Architect A person licensed to practice the profession of architecture under the education law of the state in which the person practices.

Area well Used around basement windows to hold back the soil; usually constructed of galvanized, ribbed steel, concrete, or masonry.

Armored/BX cable A cable with a flexible metal covering that is often used with appliances.

As-built drawings Construction drawings that have been revised to show changes made to the original plans during construction.

Askarel A generic term for a group of nonflammable synthetic chlorinated hydrocarbons used as electrical insulating media. Askarels of various compositional types are used. Under arcing conditions, the gases produced, while consisting predominantly of noncombustible hydrogen chloride, can include varying amounts of combustible gases, depending on the askarel type.

Aspect ratio The ratio of height to width of a shear wall.

Assessment A tax levied on a property, or a value placed on the worth of a property.

Asset In the technical terminology of accounting, assets are the resources with which a business operates (e.g., buildings, equipment, tools, and cash).

Asset management Managing assets so as to maximize value for the owner.

Asset management The management of real property, installed equipment and furniture, furnishings, and equipment using computers for scheduling work assignments, monitoring equipment usage, controlling inventory and the like.

Assumption Allows a buyer to assume responsibility for an existing loan instead of getting a new loan.

Astragal A molding that is attached to one of a pair of swinging double doors, against which the other door strikes.

Attached drawer A type of drawer in which the drawer face is attached to the front of a self-contained drawer box. See **integral drawer**.

Attachment plug (plug cap or plug) A device that, by insertion in a receptacle, establishes a connection between the conductors of the attached flexible cord and the conductors connected permanently to the receptacle.

Attic access An opening that is placed in the ceiling of a home to provide access to the attic.

Attic ventilators In houses, screened openings provided to ventilate an attic space.

Auger A tool that includes a screw-shaped shaft that digs a hole when turned.

Aught American standard gauge unit of measure for wire sizes that are larger than 2 gauge. See **American standard gauge**.

Authority having jurisdiction The organization, office, or individual responsible for approving equipment, materials, and installation, or a procedure.[2]

Automatic Self-acting; operating by its own mechanism when actuated by some impersonal influence, as, for example, a change in current, pressure, temperature, or mechanical configuration.

Awning window A window unit that opens by moving the bottom of the window sash outward. The top of the window sash is attached with hinges.

Back charge Billings for work performed or costs incurred by one party that, in accordance with the agreement, should have been performed or incurred by the party to whom billed. Owners bill back charges to general contractors, and general contractors bill back charges to subcontractors. Examples of back charges include charges for cleanup work or to repair something damaged by another subcontractor.

Backer board A board not part of the original framing that is placed in a wall or joist system to provide backing for the attachment of drywall, sheathing, or an intersecting wall.

---

2. The phrase "authority having jurisdiction" is used in National Fire Protection Agency (NFPA) documents in a broad manner since jurisdictions and approval agencies vary, as do their responsibilities. Where public safety is primary, the authority having jurisdiction may be a federal, state, local, or other regional department or individual such as a fire chief; fire marshal; chief of a fire prevention bureau, labor department, or health department; building official; electrical inspector; or others having statutory authority. For insurance purposes, an insurance inspection department, rating bureau, or other insurance company representative may be the authority having jurisdiction. In many circumstances, the property owner or his or her designated agent assumes the role of the authority having jurisdiction; at government installations, the commanding officer or departmental official may be the authority having jurisdiction.

Backfill The replacement of excavated earth into a trench around or against a basement or crawl space foundation wall.

Backflow (drainage) Reversals of drainage flow.

Backflow preventer A device or means to prevent backflow.

Backing A framing member installed at a nonlayout position so that other framing members can be securely attached later in the construction process. Carpet backing holds the pile fabric in place.

Backing layer (vinyl flooring) One of three layers of material typically found in vinyl floor covering. The backing layer is the bottom layer. See **pattern layer** and **perimeter backing**.

Backloaded insulation Thermal/acoustical insulation that is placed above the ceiling suspension system and laid across the horizontal grid members above the acoustical panels or tile. Also referred to as backloading.

Backout The work the framing contractor does after the mechanical subcontractors (heating, plumbing, electrical) finish their phase of work at the rough (before insulation) stage to prepare for a municipal frame inspection. Generally, the framing contractor repairs anything disturbed by others and completes all framing necessary to pass a rough frame inspection.

Back pressure Pressure that is greater than the water supply pressure and, therefore, potentially causes a backflow.

Backsiphonage The flowing back of contaminated water into potable water due to negative pressure.

Backsplash The part that fits on the wall behind the countertop and is designed to protect the wall from countertop and sink splashes. The backsplash is often made from the same materials as the countertop. See **block backsplash** and **full backsplash**.

Baffle A deflector used to stop the transmission of a material such as sound, light, or a liquid. A noncombustible baffle should be placed around a recessed light fixture to prevent heat produced by the light from being trapped by insulation where it can build up and cause a fire.

Baked-enamel finish The application of special enamel paint onto a nonporous surface by baking it onto a surface, usually metal.

Ballast Smooth aggregate placed on the surface of a roof to weigh down the roofing; also protects the roof materials from ultraviolet light.

Ballast A transformer that steps up the voltage in a florescent lamp.

Balloon-framed wall Framed walls (generally over 10 ft. tall) that run the entire vertical length from the floor sill plate to the roof. This is done to eliminate the need for a gable-end truss.

Balloon loan A loan that has a series of monthly payments with the remaining amount due in a large lump sum payment at the end.

Baluster One of a series of vertical posts that are placed at regular intervals along the length a balustrade. Balusters are typically similar in pattern to the Newel post but are much smaller in diameter or thickness. Balusters are attached to a top rail and to a bottom rail or directly to a stair tread or floor. See **Balustrade**.

Balustrade Railing system found on stairs or along open areas between floors. Most balustrades contain newel posts, balusters, and top rails. Some balustrades contain bottom rails. See **baluster, bottom rail, bread loaf top rail, false tread and riser, fillet, goose neck, newel post, one-quarter turn, rosette, skirt, stair bracket, top rail**, and **volute**.

Band joist A piece of lumber to which the ends of the joists are nailed or screwed. A band joist is critical to the strength of the floor system because it holds the regular joist ends in their vertical position.

Barge A horizontal beam rafter that supports shorter rafters.

Barge board A decorative board covering the projecting rafter (fly rafter) of the gable end. At the cornice, this member is a fascia board.

Base block A square or rectangular piece that is placed between the bottom of the casing and the floor; also called plinth block. See **corner block**.

Baseboard See **base molding**.

Base coat First coat of any finish substance on a surface, such as the first coat of synthetic stucco applied on the stucco sheathing over a wire or glass fiber mesh.

Base flashings A continuation of a built-up roofing membrane, at the upturned edges of the watertight tray. They are normally made of bitumen-impregnated, plastic, or other non-metallic materials and applied in an operation separate from the application of the membrane itself. See **counterflashings**.

Baseline irrigation water use The amount of water that would be used by a typical method of irrigation for the region.

Basement window inserts The window frame and glass unit that is installed in the window buck.

Base molding A molding that covers the corner between the floor and wall.

Base sheet The first layer of a multiple ply membrane roof system. Installed by rolling strips of special base sheet-style roll roofing over the roof deck and nailing the base sheet to the roof deck. Once the base sheet is in place, hot tar is mopped over it.

Base shoe A molding placed at the corner between the base molding and floor. Usually used when a wood-finish floor is installed.

Base shoe Molding used next to the floor on interior baseboard. Sometimes called a carpet strip.

Bat A half-brick.

Bathroom An area including a basin with one or more of the following: a toilet, a tub, or a shower.

Batt A section of fiberglass or rock wool insulation measuring 15 or 23 in. wide by 4 to 8 ft. long and varying in thickness. Sometimes "faced" (meaning to have a paper covering on one side) or "unfaced" (without paper).

Batten-type standing seam A standing seam roof in which the panels are raised up and fastened together

and then a batten strip is placed over the seam to form a watertight seal.

Bay window A window unit installed in an area that projects out from the wall. The exterior wall typically forms a 45-degree angle on each end of the bay window area.

Beam A structural member transversely supporting a load; a structural member carrying building loads (weight) from one support to another; sometimes called a girder.

Beam pocket A notch or opening at the top of a bearing wall or supporting column that secures and bears the weight of a beam.

Bearing header A beam placed perpendicular to joists and to which joists are nailed in framing for a chimney, stairway, or other opening.

Bearing header A wood lintel.

Bearing header The horizontal structural member over an opening (e.g., over a door or window).

Bearing partition A partition that supports any vertical load in addition to its own weight.

Bearing point A point where a bearing or structural weight is concentrated and transferred to the foundation.

Bearing points The location or point on a member where another member supports it. Typically, a truss has two bearing points: one at each exterior wall.

Bearing wall A wall that supports any vertical load in addition to its own weight.

Bearing wall Any wall that supports a load above it, such as a roof system or a floor system. A bearing wall is a structural part.

Bedrock A subsurface layer of earth that is suitable to support a structure.

Berber carpet A style of carpet with a distinctive, short looped pile. See **indoor-outdoor carpet**, **nylon carpet**, **sculptured carpet**, **shag carpet**, and **wool carpet**.

Beveled A cut at a non–right angle to the main surface, forming a sloping surface.

Bid A formal offer by a contractor, in accordance with plans, specifications, terms, and conditions, to do a project in all or in part at a certain price.

Bid bond A bond issued by a surety on behalf of a contractor that provides assurance to the recipient of the contractor's bid that, if the bid is accepted, the contractor will execute a contract and provide a performance bond. Under the bond, the surety is obligated to pay the recipient of the bid the difference between the contractor's bid and the bid of the next lowest responsible bidder if the bid is accepted and the contractor fails

to execute a contract or to provide a performance bond.

Bidding requirements The procedures and conditions for the submission of bids. The requirements are included in documents, such as the notice to bidders, advertisements for bids, instructions to bidders, invitations to bid, and sample bid forms.

Bid security Funds or a bid bond submitted with a bid as a guarantee to the recipient of the bid that the contractor, if awarded the contract, will execute the contract in accordance with the bidding requirements of the contract documents.

Bid shopping A practice by which contractors, both before and after their bids are submitted, attempt to obtain prices from potential subcontractors and material suppliers that are lower than the contractors' original estimates on which their bids are based or, after a contract is awarded, a practice by which they seek to induce subcontractors to reduce the subcontract price included in the bid. Bid shopping is considered to be unethical.

Binder A receipt for a deposit to secure the right to purchase a home at agreed terms by a buyer and seller.

Biofuel-based electrical systems Electrical power systems that run on renewable fuels derived from organic materials, such as wood by-products and agricultural waste.

Biological control The use of chemical or physical water treatments to inhibit bacterial growth in cooling towers.

Biomass Plant material from trees, grasses, or crops that can be converted to heat energy to produce electricity.

Bird stop Material used to fill the space under the first course of tile at the eave line to prevent birds from nesting in the roofing.

Biscuit Joint between two boards made by using a biscuit saw to notch out the ends of the joined boards. A premanufactured biscuit fits into the slots made by the biscuit saw. The glued biscuit swells as the glue soaks in, forming a very tight fit when the joint dries.

Biscuit saw A special saw used to cut a notch in boards that will be joined with a biscuit joint.

Black label shingles Utility grade shingles with sapwood, flat grain, and large knots; commonly used on garages and barns.

Blackwater Does not have a single definition. Wastewater from toilets and urinals is always deemed blackwater. Wastewater from kitchen sinks, showers, and bathtubs may also be considered blackwater. You should check your local codes for the definition used in your area.

Blankets Fiberglass or rock wool insulation that comes in long rolls 15 or 23 in. wide.

Bleed-off or blowdown The release of built-up solids in a cooling tower, accomplished by removing a portion of the concentrated recirculating water that carries dissolved solids.

Bleed-off rate The frequency with which the dissolved minerals and dirt are removed from the cooling tower. It varies depending on the mineral content and scaling tendency of the entering water.

Blind miter A joint made by butting the first piece of material into a corner and then shaping the second piece so that it conforms to the outline of the first piece; also called a cope joint.

Blind nailing Driving a nail into a part of the board that will not be visible on the finished product. See also **face nailing**.

Block A masonry product that is used in the assembly of footings, foundation walls, along with both interior and exterior walls. Blocks are precast to specific dimensions and are available in many shapes and styles.

Block backsplash A thick, relatively short backsplash with a square top. See **backsplash** and **full backsplash**.

Block building system Involves using block to construct the perimeter foundation system and the exterior bearing walls. The footings in a block building system may also be built from block but are often made from concrete.

Block cell Refers to the open cavities often found in blocks. These cells may be filled with either grout or insulation.

Blocked (door blocking) Wood shims used between the door frame and the vertical structural wall framing members.

Blocked (rafters) Short two-by-fours used to keep rafters from twisting; installed at the ends and at midspan.

Blocking Short pieces of material used to provide solid bridging over bearing points and to block fire from quickly spreading into other parts of the framing; usually found in joists over every bearing wall or beam and in studs at every connection with stair stringers or dropped ceilings.

Block out To install a box or barrier within a foundation wall to prevent the concrete from entering an area. For example, foundation walls are sometimes "blocked" in order for mechanical pipes to pass through the wall, to install a crawl space door, and

to depress the concrete at a garage door location.

Blown-in insulation Insulation that is broken down by a machine that blows it into place. Blown-in insulation is often installed above the ceiling line and in wall cavities.

Blue label shingles Shingles that are made from the highest quality heartwood and clear cedar with 100% edge grain. Most residential structures use red or blue label shingles.

Blueprint(s) A type of copying method often used for architectural drawings. Usually used to describe the drawing of a structure that is prepared by an architect or designer for the purpose of design and planning, estimating, securing permits, and actual construction.

Blue stake Another phrase for utility notification. This is when a utility company (telephone, gas, electric, cable TV, sewer and water, etc.) comes to the jobsite and locates and spray paints the ground and installs little flags to show where their service is located underground.

Board and batten siding Vertical siding in which boards are installed first with small spaces between them. Narrower boards, called battens, are then installed over the small spaces.

Board foot Lumber quantities are expressed in ft., board measure (BM or ft. bm.) or in board ft. (bd. ft.), or in thousand board ft. (M bd. ft.). One board ft. is the amount of lumber in a rough-sawed board 1-ft. long, 1-ft. wide, and 1-in. thick (144 cu. in.) or the equivalent volume in any other shape, for example, 1-ft. long by 6-in. wide by 2-in. thick. The originals or "nominal" dimensions and volumes determine the number of board ft. in a quantity of dressed lumber. The process of surfacing or other machining does not lessen the board ft. that are sold. Examples: 1 in. × 12 in. × 16 ft. = 16 board ft., 2 in. × 12 in. × 16 ft. = 32 board ft.

Board-on-board siding Vertical siding installed with gaps between boards. Boards of the same size are then installed over the gaps.

Board siding Siding made from wood, hardboard, or pressed wood by-products; usually installed horizontally, one board at a time.

Bond Siding made from wood, hardboard, or pressed wood byproducts; usually installed horizontally, one board at a time.

Bond beam Horizontal beam poured inside the U-block for reinforcement of block walls. A bond beam is made by filling the block cells with either grout or insulation up to the level of the bottom of the U-block. Reinforcing steel is placed, and the U-block is filled with grout.

Bonding (bonded) The permanent joining of metallic parts to form an electrically conductive path that ensures electrical continuity and the capacity to conduct any current likely to be imposed safely.

Bonding jumper A conductor that ensures that the required electrical conductivity between connected metal raceways is both safe and reliable. (This is not the same as grounding, but bonding jumpers are essential components of the bonding system that is an essential component of the grounding system. Note: the NEC does not authorize the use of the earth as a bonding jumper—that's because the resistance of the earth is more than 100,000 times greater than that of a bonding jumper.)

Bond or bonding An amount of money (usually $5000–$10,000) deposited with a governmental agency to secure a contractor's license. The bond may be used to pay for the unpaid bills or disputed work of the contractor.

Bond or bonding You may be required to provide any or all of the following bonds:

- Performance bonds guarantee the faithful performance of all work required to complete the contract. Payment bonds guarantee the payment of all bills for labor and materials used in the work, including materials purchased for the project but not included in it.

- Bid bonds guarantee that the contractor, upon being declared the successful bidder, will enter into a contract with the owner for the amount of the submitted bid and will provide contract bonds as required.

- A maintenance bond guarantees the performance of the contract. It covers the time after the building is completed. This bond stipulates the building stability and maintenance responsibility of the construction company after the building has been completed.

- Permit bonds are required by some local authorities to indemnify the authority against any costs arising from the contractor using public streets, crossing curbs, connecting to public utility lines, etc.

- Supply bonds guarantee the quality and quantity supplied to the owner. These are rarely used on construction projects.

Booking The process of folding the pasted side of wallpaper over onto itself. Booking allows the glue to cure without drying unevenly.

Book-matched veneer A veneer pattern produced by turning over every other veneer strip. On a surface, the strips look much like mirror images of each other. See **slip-matched veneer**, **unmatched veneer**, **veneer**, and **whole piece veneer**.

Boom A truck used to hoist heavy material up and into place; to put trusses on a home or to set a heavy beam into place.

Booster tile Small roof tile placed under the cap tiles on the starter course only.

Border cut Referring to ceilings, a cut made on both the ceiling panel and grid at the perimeter.

Bottom chord Lower or bottom member of a truss.

Bottom plate The two-by-fours or -sixes that lay on the subfloor upon which the vertical studs are installed; also called the sole plate.

Bottom rail The lower rail of a balustrade into which the bottom ends of balusters connect. See **balustrade** and **top rail**.

Bowstring truss A truss with a curved top chord and horizontal bottom chord so that the top looks like a bow string and the bottom looks like a bow.

Bow window A window unit that projects out from the wall in an arch.

Box (cabinet) The storage section of the cabinet. See **face** and **face frame**.

Box beam Made from steel or wood, they are formed like a long box with four sides and are hollow in the center.

Box nails Framing fasteners with a slightly smaller shank or shaft than common nails, but with the same length and the same size heads as comparable common nails. Box nails are less likely to split the wood than common nails but they are not quite as strong. See **common nails, gun nails**, and **sinkers**.

Brace An inclined piece of framing lumber applied to wall or floor to strengthen the structure. Often used on walls as temporary bracing until framing has been completed.

Branch circuit The circuit conductors between the final overcurrent device protecting the circuit and the outlet(s).

Branch circuit (appliance) A branch circuit that supplies energy to one or more outlets to which appliances are to be connected and that has no permanently connected luminaries (lighting fixtures) that are not part of an appliance.

Branch circuit (general-purpose) A branch circuit that supplies two or more receptacles or outlets for lighting and appliances.

Branch circuit (individual) An individual branch circuit, as defined in Article 100 of the National Electric Code (NEC), "is a branch circuit that supplies only one utilization equipment." Although a duplex receptacle is installed and mounted by one strap or yoke, it is considered two receptacles. A branch circuit supplying only a duplex receptacle and no other device is not an individual branch circuit.

Branch circuit (multiwire) A branch circuit that consists of two or more ungrounded conductors that have a voltage between them, and a grounded conductor that has equal voltage between it and each ungrounded conductor of the circuit and that is connected to the neutral or grounded conductor of the system.

Bread loaf top rail A common type of top rail that has a profile shaped like a loaf of bread. See **balustrade** and **top rail**.

Break (elevator) a spring-loaded clamping device that prevents the elevator from moving when the car is at rest and no power is applied to the hoistway motor.

Break drum (elevator) A round, machined surface on the motor shaft that the break clamps for stopping.

Breaker panel The electrical box that distributes electric power entering the home to each branch circuit (each plug and switch), composed of circuit breakers.

Brick ledge Part of the foundation wall where brick (veneer) will rest.

Brick lintel The metal angle iron that bricks rest on, especially above a window, door, or other opening.

Brick mold Trim used around an exterior doorjamb to which siding butts.

Brick tie A small, corrugated metal strip at 1 in. × 6 in. to 8 in. long nailed to wall sheeting or studs. They are inserted into the grout mortar joint of the veneer brick, and hold the veneer wall to the sheeted wall behind it.

Brick veneer A vertical facing of brick laid against and fastened to sheathing of a framed wall or tile wall construction.

Bridging Small wood or metal members that are inserted in a diagonal position between the floor joists or rafters at midspan for the purpose of bracing the joists/rafters and spreading the load.

Broom finish The most common exterior flatwork finish; a slightly rough texture achieved by running a broom over freshly troweled concrete.

Brown coat Second coat of stucco, applied over the scratch coat. The purpose of the brown coat is to provide a relatively smooth surface for the finish coat. The brown coat is troweled over the scratch coat and then smoothed with a long float. See also **finish coat** and **scratch coat**.

Bruised composition shingles A composition shingle that has been permanently dented by a hailstone but has not fractured. See also **fractured composition shingles** and **granular loss**.

Brush texture Finish applied to drywall with a brush.

BTU (British thermal unit) The amount of heat required to raise the temperature of one pound of water one degree Fahrenheit.

Buck Often used in reference to rough frame opening members (e.g., door bucks are used in reference to metal doorframes). See **window bucks**.

Buffer A device designed to stop a descending elevator car or counterweight beyond its normal limit and to soften the force with which the elevator runs into the pit during an emergency.

Buffer channel A channel in the pit floor of a traction elevator that supports buffers and guide rails.

Buffer springs Large diameter springs that are permanently placed in a traction elevator pit for the purpose of stopping a descending car or counterweight beyond its normal limit of travel. The distance that the springs compress is called a buffer stroke.

Builder's risk insurance Insurance coverage taken on a specific construction project for the period of construction. Some contracts require this insurance to protect the owner.

Building A, usually, roofed and walled structure built for permanent use (as for a dwelling). Buildings stand alone or are cut off from adjoining structures by firewalls with all openings between structures protected by approved fire doors.

Building automation system (BAS) A computer-based monitoring and control system that coordinates, organizes, and optimizes building control subsystems, including lighting and equipment scheduling and alarm reporting.

Building codes Community ordinances governing the manner in which a facility may be constructed or modified.

Building footprint The area of the site occupied by the building structure, not including parking lots, landscapes, and other nonbuilding facilities.

Building inspector An individual trained and certified to inspect completed construction work for compliance with accepted building codes and ordinances.

Building insurance Building insurance covers the structure.

Building paper A general term for papers, felts, and similar sheet materials used in buildings without reference to their properties or uses. Generally comes in long rolls.

Building permit A permit issued by the local government, usually the county or city, after a fee has been paid and plans have been reviewed and approved. Normally construction cannot begin until after the permit is issued nor will a certificate of occupancy be issued until a building inspector has approved all work.

Built-in cabinets Cabinets that are hand-built on site. See **custom cabinets**, **mass-produced cabinets**, and **milled cabinets**.

Built-up roof See **multiple-ply membrane**.

Bull float A tool with a long handle that slides on the surface of concrete to press the course aggregate down and to raise the cream.

Bull nose (drywall) Rounded drywall corners.

Bundle A package of shingles. Normally, there are three bundles per square and 27 shingles per bundle.

Bunk A unit of lumber consisting of several pieces or sticks that are banded together for convenience in shipping and delivery.

Burn pattern The direction a fire burns. Fire burns up and away from the point of origin toward an oxygen source.

Butt edge The lower edge of the shingle tabs.

Butterfly roof A roof style consisting of two planes that slope inward forming a V, lower in the center than at the outside edge.

Butt hinge A hinge made up of two flat, rectangular plates with a pin connecting them.

Butt joint A joint made by placing two square-cut pieces of material end to end without any overlap.

Butt seam The seam found when two members are joined together end-to-end without overlapping the members.

Buy down A subsidy (usually paid by a builder or developer) to reduce monthly payments on a mortgage.

By-fold door A door with two slabs that are connected to each other with hinges. When closed, the slab ends butt against each other. When opened, the slabs fold onto each other. A track at the top of the by-fold door holds the slabs in position. Sometimes spelled bifold.

Bypass doors Door with two flush slabs that are mounted on tracks. Each door part slides parallel to the other. Bypass doors are common on closets and patios. Also called sliding doors. Sometimes spelled bipass.

Cabinet An enclosure that is designed for either surface mounting or flush mounting and is provided with a frame, mat, or trim in which swinging doors are or can be hung.

Cables Ropes, usually four to six in number, used to support an elevator car. Ropes pass over the drive sheave to the counterweight, either pulling up the car or lowering it. The amount of the drive sheave actually in contact with the cable is called cable wrap.

Caisson A 10-in. or 12-in. diameter hole drilled into the earth and embedded into bedrock 3 to 4 ft. A caisson serves as the structural support for a type of foundation, wall, porch, patio, monopost, or other structure. Usually, reinforcing bars (rebar) are inserted into and run the full length of the hole and concrete is poured into the caisson hole.

Calrod burner An electrical burner made from coiled steel. See **ceran top** and **halogen burner**.

Camber Refers to the slight bow or arch that is found in many building materials. Camber is sometimes used as a synonym for crown. However, crown usually refers to the natural distortion that occurs in lumber whereas camber usually refers to a built-in bow that was engineered by the manufacturer. See **crown**.

Can Housing or container for a recessed light unit. The can is installed during electrical rough-in.

Cantilever A beam or joist with an end portion that hangs out past the structural part that supports it (e.g., a diving board is cantilevered).

Cant strip A beveled piece of material placed where the roofing material turns up such as at the intersection of a parapet wall and the roof deck. It is used to soften the angle that must be covered by the roofing membrane.

Cap The upper member of a column, pilaster, door cornice, molding, or fireplace.

Cap flashing The portion of the flashing attached to a vertical surface to prevent water from migrating behind the base flashing.

Capital The principal part of a loan (i.e., the original amount borrowed).

Capital and interest A repayment loan and the most conventional form of home loan. The borrower pays an amount each month to cover the amount borrowed (or capital or principal) plus the interest charged on capital.

Capped rate The mortgage interest rate will not exceed a specified value during a certain period of time, but it will fluctuate up and down below that level.

Cap row Top course of shingles that does not have another course overlapping it. There is a cap row on the ridge of the roof.

Cap sheet The top layer of multiple-ply membrane roof system usually covered with one of three finishes: (1) a smooth, flood coat that is painted to prevent sun damage; (2) a flood coat with aggregate covering; or (3) a cap sheet with mineral granules embedded into the surface. See **flood coat**.

Cap tile A U-shaped roofing tile that forms the peaks in a barrel tile roof. Cap tiles are placed at the intersection of pan tiles with the U facing down. Water drains off of the peaks formed by the cap tiles and into the pan tiles. Specially designed cap tiles are also used on the rake, ridges, and hips.

Carbon dioxide (CO_2) levels An indicator of ventilation effectiveness inside buildings. CO_2 concentrations greater than 530 parts per million (ppm) above outdoor CO_2 conditions generally indicate inadequate ventilation. Absolute concentrations of CO_2 greater than 800 to 1000 ppm generally indicate poor air quality for breathing.

Car operating panel (car station) A panel mounted in the elevator car containing the car operating controls, such as call register (floor) buttons, door open and close, alarm, emergency stop, and any other buttons or keyswitches that may be required for operation.

Carpet cove Carpet that wraps a small distance up the wall.

Carpet grain Direction in which the carpet fibers slant.

Casement Frames of wood or metal enclosing part (or all) of a window sash. May be opened by means of hinges affixed to the vertical edges.

Casement window A window unit that opens by swinging the side of the window sash outward like a door. The opposite side of the window sash is attached with hinges.

Casing Framing that surrounds a door, covering the space between the jamb and the wall surface.

Catwalk A support member attached near the center of the bottom chord of a truss system to hold the trusses in a vertical position.

Caulking A flexible material used to seal a gap between two surfaces (e.g., between pieces of siding or the corners in tub walls).

Caulking To fill a joint with mastic or asphalt plastic cement to prevent leaks.

CCA (chromated copper arsenate) A pesticide that is forced into wood under high pressure to protect it from termites, other wood boring insects, and decay caused by fungus.

Ceiling attenuation system Rates a ceiling's efficiency as a barrier to airborne sound transmission between adjacent rooms. A single-figure rating derived from the normalized ceiling attenuation values in accordance with classification ASTM E 413, except that the resultant rating shall be designated ceiling attenuation class (defined in ASTM E 1414). Previously expressed as CSTC (ceiling sound transmission class).

Ceiling duty classification The load-carrying capability of grid main beams (per ASTM C635) pounds per linear ft. (light: 5 lbs.; intermediate: 12 lbs.; heavy: 16 lbs.).

Ceiling joist One of a series of parallel framing members used to support ceiling loads and supported in turn by larger beams, girders, or bearing walls. Also called roof joists.

Ceiling suspension system A system of metal members that are designed to support a suspended ceiling. Ceiling suspension systems are usually designed to accommodate lighting fixtures and diffusers.

Cellulose insulation Insulation that is made from shredded paper. Cellulose insulation should be treated with a fire retardant chemical, or it may create a fire hazard.

Celotex Black fibrous board that is used as exterior sheathing.

Cement The gray powder that is the "glue" in concrete (and should not be confused with concrete); also, any adhesive.

Cement board underlayment An underlayment made from cement board that has a smooth finish for use beneath vinyl floors. Cement board underlayment is almost impervious to water damage. See **gypsum-based underlayment, lauan plywood underlayment, particleboard underlayment, plywood underlayment,** and **untempered hardboard underlayment**.

Cementuous board A type of cement board attached to a substrate to create an isolation membrane. The joints between the boards are then filled and leveled. See **isolation membrane, mortar bed,** and **thin-set tile**.

Center bearing wall A wall on the interior of a structure that is built to support the weight of the floor system above it. The center bearing wall is usually constructed along the center line of the structure. This is a structural part.

Ceramic mosaic tile Small tiles, usually 1 in. × 1 in., 1 in. × 2 in., or 2 in. × 2 in., that are typically made from porcelain or natural clay; generally used in bathtub and shower enclosures and on counter tops. See **glazed wall tile** and **quarry tile**.

Ceran top A brand name for a type of specialty glass used in flat-surface cooktops. The burners are placed under the glass and cooking utensils are placed on top of the glass.

CFM (cubic feet per minute) A rating that expresses the amount of air a blower or fan can move. The volume of air (measured in cu. ft.) that can pass through an opening in 1 minute.

Chain of custody A tracking procedure for documenting the status of a product from the point of harvest or extraction to the ultimate consumer end use, including all successive stages of processing, transformation, manufacturing, and distribution. Establishing chains of custody is important when LEED certification is desired. Doing so can add to your costs.

Chair rail Interior trim material installed about 3 to 4 ft. up the wall, horizontally.

Chalk line A line made by snapping a taut string or cord dusted with chalk. Used for alignment purposes.

Change order A written document that modifies the plans, specifications, or the price of the construction contract.

Charring As wood burns, the areas of the wood being consumed will begin to char. Charred portions of wood are not structurally stable. When the surface of structural framing members is charred more than 1/8 in., the framing member will generally have to be replaced.

Chase A framed, enclosed space around a flue pipe or a channel in a wall or through a ceiling for something to lie in or pass through.

Chemical runoff Water that transports chemicals from the building landscape, as well as from surrounding streets and parking lots, to rivers and lakes. Runoff chemicals may include gasoline, oil, antifreeze, and salts.

Chemical treatment The use of biocidal, conditioning, dispersant, and scale-inhibiting chemicals to control biological growth, scale, and corrosion in cooling towers. Alternatives to conventional chemical treatment include ozonation, ionization, and UV light.

Chem sponge Also called a dry sponge or chemical sponge, a chem sponge is treated with special chemicals for use in removing soot from walls. Neither water nor other liquids are used with a chem sponge.

Chimney cricket A small roof built behind the chimney to move precipitation around the chimney and off the roof; also called a chimney saddle.

Chink To install fiberglass insulation around all exterior door and window frames, wall corners, and small gaps in the exterior wall.

Chip board A manufactured wood panel made out of 1- to 2-in. wood chips and glue; often used as a substitute for plywood in the exterior wall and roof sheathing; also called OSB (oriented strand board) or wafer board.

Chlorofluorocarbons (CFCs) Hydrocarbons that are used as refrigerants and cause depletion of the stratospheric ozone layer.

Cinder block A masonry unit made from Portland cement and cinder. Cinder blocks are lighter and have better insulation qualities than concrete masonry units.

Circle A shape in which all points on the perimeter are the same distance from the center. The formula for calculating the area of a circle is: pi × radius^2 = area. The formula for the perimeter is: pi × diameter = circumference.

Circuit The path of electrical flow from a power source through an outlet and back to ground.

Circuit breaker A device that looks like a switch and is usually located inside the electrical breaker panel or circuit breaker box. It is designed to (a) shut off the power to portions of or the entire house and (b) to limit the amount of power flowing through a circuit (measured in amperes). The 110-volt household circuits require a fuse or circuit breaker with a rating of 15 or a maximum of 20 amps. 220-volt circuits may be designed for higher amperage loads (e.g., a hot water heater may be designed for a 30 amp load and would, therefore, need a 30 amp fuse or breaker).

Circular saw A power tool with a circular blade that rotates at a high speed so that the teeth on the blade will cut the material, usually wood.

Circular stair A stair system that winds in a curving pattern, usually, but not always around a common center.

Circumference The length of the perimeter of a circle, calculated with the formula pi × diameter = circumference; can be used interchangeably with perimeter.

Class "A" The optimum fire rating issued by Underwriters Laboratories on roofing. The building codes in some areas require this type of roofing for fire safety.

Class "C" The minimum fire rating issued by the Underwriters Laboratories for roofing materials.

Clay block A masonry unit made from clay. Most often used on commercial structures; also commonly called atlas block.

Clay tile Tile made from clay that has been forced through an extruder, cut to size, air dried, and then fired in a kiln.

Clean out An opening that provides access to a drain line. Clean outs are closed with threaded plugs.

Clear all heart High-grade redwood lumber that is free of knots, pitch, and blemishes. The grain of clear all heart is usually fairly straight.

Clerestory roof A roof style consisting of two sides that slope in opposite directions with a vertical wall section extending between the peaks. The vertical wall contains windows that provide light and ventilation into the building. The clerestory roof is common on condominium and passive solar homes.

Clip ties Sharp, cut metal wires that once held the foundation form panels in place and protrude out of a concrete foundation wall.

Closed valley The system intersection of two roof surfaces where the courses of shingles meet and cover the valley flashing. See also **half-laced valley**, and **laced valley**.

Clustered wiring When wires in a group are each covered with insulation (except in some cases for the copper ground) and then the entire cluster of wires is also covered by plastic insulation.

Coal tar pitch A hydrocarbon substance created by processing coal; may be used to waterproof membrane roofing.

CO and CofO An abbreviation for "certificate of occupancy." This certificate is issued by the local municipality and is required before anyone can occupy and live within the home.

It is issued only after the local municipality has made all inspections and all monies and fees have been paid.

Cold-air return The ductwork (and related grills) that carries room air back to the furnace for reheating.

Cold joint The joint that occurs when a batch of fresh concrete is placed next to concrete that is set or concrete that is less plastic with no vibration or rodding to cause the two batches to consolidate.

Collar Preformed flange placed over a vent pipe to seal the roofing above the vent pipe opening; also called a vent sleeve.

Collar beam Nominal 1- or 2-in.-thick members connecting opposite roof rafters. They serve to stiffen the roof structure.

Collar tie Horizontal member that is used to tie rafters together above the top plate. The collar tie strengthens the roof member and may be used for fastening the ceiling; may also be referred to as a collar beam.

Colonial base and casing A commonly used molding pattern.

Colonist door Brand name of door that is made of pressed wood fiber and has numerous raised panels.

Column A vertical structural compression member that supports loads.

Combustion air The ductwork installed to bring fresh, outside air to the furnace or hot water heater. Normally two separate supplies of air are brought in: one high and one low.

Combustion chamber The part of a boiler, furnace, or woodstove where the burn occurs; normally lined with firebrick or molded or sprayed insulation.

Comfort criteria Specific original design conditions that include temperature (air, radiant, and surface), humidity, air speed, outdoor temperature design conditions, outdoor humidity design conditions, clothing (seasonal), and expected activity (ASHRAE Standard 55-2004).

Commissioning cycle A quality-oriented process for achieving, verifying, and documenting that the performance of facilities, systems, and assemblies meets defined objectives and criteria. It is an inclusive process for planning, delivering, verifying, and managing risks of and to a facilities' critical functions.

Common gable truss A truss used to make a gable roof. It usually spans from outside wall to outside wall without relying on interior bearing walls for support. All trusses in a gable truss roof will be common gable trusses except the last truss on each end that are gable end trusses.

Common nails The standard fasteners used by framers. The following are typical common nail sizes:

Size (penny)	Shank diameter (nominal)	Shank length (in.; nominal)
2d	0.072	1
3d	0.083	1¼
4d	0.109	1½
5d	0.109	1¾
6d	0.120	2
8d	0.134	2½
10d	0.148	3
12d	0.148	3¾
16d	0.165	3½
20d	0.203	4
30d	0.220	4½
40d	0.238	5
60d	0.238	6

See **box nails** and **sinkers**.

Common rafter A full-length rafter that extends from the top wall plate to the ridge.

Compactable fill Soil that is capable of being compacted so as to provide a solid substance under the structural parts. Compactable fill is almost always placed in layers (also called lifts) that are thin enough (usually 4 to 6 in.) to allow the compaction device to be effective throughout the layer. Each layer is thoroughly compacted before the next one is placed.

Completed-contract method An accounting method that recognizes revenues and gross profit only when the contract is completed.

Com-ply Type of wood product used for sheathing. It consists of two thin layers or veneers of wood on the outside and in the middle has a wood flake and resin center.

Composite wood Man-made wood; includes a range of derivative wood products that are manufactured by binding together the strands, particles, fibers, or veneers of wood, together with adhesives, to form composite materials.

Composition shingle A shingle made from an organic or fiberglass mat that is saturated with asphalt or coal tar pitch. Granules are embedded in the surface that is exposed to the weather; also commonly called organic, asphalt, or fiberglass shingles.

Composting toilet systems Sometimes called biological toilets, dry toilets, and waterless toilets, they contain and control the composting of excrement, toilet paper, carbon additive, and, optionally, food wastes. Unlike a septic system, a composting toilet system relies on unsaturated conditions (material cannot be fully immersed in water), where aerobic bacteria and fungi break down wastes, just as they do in a yard waste composter.

Compressed workweek The rearrangement of the standard workweek, 5 consecutive 8-hour days, by increasing the daily hours and decreasing the number of days in the workweek (e.g., 10-hour days for 4 days per week or 9-hour days for 9 of 10 consecutive days). May save on overtime if, say, a project includes work that cannot be completed in 8 hours but can be completed in one 10-hour shift.

Compression Crushing force. Compression is the opposite of tension.

Compression fitting A fitting used on all types of pipe where a ferrule or a gasket is compressed against the fitting by tightening a threaded nut.

Compression web A member of a truss system that connects the bottom and top chords and provides downward support.

Compressor A mechanical device that pressurizes a gas in order to turn it into a liquid, thereby allowing heat to be removed or added. A compressor is the main component of conventional heat pumps and air conditioners. In an air-conditioning system, the compressor normally sits outside and has a large fan to remove heat.

Concealed A place rendered inaccessible by the structure or finish of the building. Wires in concealed raceways are considered concealed, even though they may become accessible by withdrawing them.

Concealed mounting system A ceiling tile suspension system that may use T-bars and splines that fit into kerfs that are cut into tile edges. Concealed mounting systems are not visible from below the ceiling. Inverted T, H and T, or Z profile grids with provisions for full plenum access are commonly incorporated into these applications.

Concrete The mixture of Portland cement, sand, gravel, and water. It is commonly reinforced with steel rods (rebar) or wire screening (mesh).

Concrete board A panel made out of concrete and fiberglass usually used as a tile backing material.

Concrete masonry block (CMU) or concrete block A masonry unit made from Portland cement and aggregate. Blocks may or may not have pigment added. CMU is the most common type of block.

Concrete stamp A form used to make patterns in concrete. After the concrete has been screed and rough finished, the stamp is pressed into the concrete. Colors may be added to make the concrete appear more natural by adding a dye admixture to the concrete or by spreading a dye on the surface.

Concrete tile Tiles made from a stiff, low slump concrete. Concrete tiles are usually heavier and less expensive than clay tiles. Concrete tiles are heavier than clay tiles and add considerable extra dead load to roofs.

Condensation Condensation is beads or drops of water (and frost in extremely cold places) that accumulates on the inside of the exterior covering of a building. Use of louvers or attic ventilators will reduce moisture condensation in attics. A vapor barrier under the gypsum lath or dry wall on exposed walls will reduce condensation.

Condensing unit The outdoor component of a cooling system. It includes a compressor and condensing coil designed to give off heat.

Conditions, covenants, and restrictions (CC and Rs) The standards that define how a property may be used and the protections the developer makes for the benefit of all owners in a subdivision.

Conduction The direct transfer of heat energy through a material.

Conductivity The rate at which heat is transmitted through a material.

Conductivity meter or EC meter A device that measures the amount of nutrients and salt in water.

Conductor Any material that carries electrical current.

Conductor (bare) A conductor that has no covering or electrical insulation.

Conductor (covered) A conductor encased within material of composition and thickness that is not recognized by the NEC as electrical insulation.

Conductor (insulated) A conductor encased within material of composition and thickness that is recognized by the NEC as electrical insulation.

Conduit A tube or pipe through which electrical wires run.

Conduit (electrical) A pipe, usually metal, in which wire is installed.

Conduit body A separate portion of a conduit or tubing system that is used to provide access to wires placed within the conduit. They are placed at a junction of two or more sections of the system or at a terminal point of the system. This differs from a junction box that allows access for pulling wires and space for splices. Conduit bodies are commonly referred to as "condulets," a term trademarked by Cooper Crouse-Hinds company, a division of Cooper Industries. Conduit bodies come in various types, moisture ratings, and materials, including galvanized steel, aluminum, and PVC.

Connector, pressure (solderless) A device that establishes a connection between two or more conductors or between one or more conductors and a terminal by means of mechanical pressure and without the use of solder. The connection to your car battery is an example of a solderless connector.

Construction contract A legal document that specifies the what, when, where, how, how much, and by whom in a construction project. A construction contract usually includes the following:

1. The contractor's registration number.
2. A statement of work quality such as "standard practices of the trades" or "according to manufacturer's specifications."
3. A set of blueprints or plans.
4. A construction timetable including starting and completion dates.
5. A set of specifications.
6. A basis for payment (e.g., fixed price or time and materials) and payment schedule.
7. Any allowances.
8. A clause that outlines how any disputes will be resolved.
9. A written warrantee.

Construction (drywall) A type of construction in which the interior wall finish is applied in a dry condition, generally in the form of sheet materials or wood paneling as contrasted to plaster.

Construction (frame) A type of construction in which the structural components are wood or depend upon a wood frame for support.

Construction and demolition (C&D) debris Waste and recyclables generated from construction, renovation, and demolition or deconstruction of preexisting structures.

Construction, demolition, and land clearing (CDL) debris Includes all construction and demolition debris plus soil, vegetation, and rocks from land clearing.

Construction indoor air quality (IAQ) management plan Measures to minimize contamination in a specific project building during construction and procedures to flush the building of contaminants prior to occupancy.

Contact adhesive A resin-based type of adhesive. The surfaces of two materials that are to be glued together are first coated with the adhesive; then the adhesive is allowed to dry (10 to 20 minutes). When the two surfaces touch each other, they adhere. See **plastic laminate countertop**.

Continuity tester A device that tells whether a circuit is capable of carrying electricity.

Continuous footing A footing design where all parts of the footing are connected. Concrete runs continuously from one section of the footing to the next with no breaks or gaps. This helps the footing resist movement during earthquakes or other types of earth movement.

Continuous load (electrical) A load where the maximum current is expected to continue for 3 hours or more.

Contractor A person or company licensed to perform certain types of construction activities. In many states, contractor and subcontractor licenses require extensive training, testing, and insurance.

Control joint Tooled, straight grooves made on concrete floors to "control" where the concrete should crack.

Controller (electrical) A device or group of devices that serves to govern, in some predetermined manner, the electrical power delivered to the apparatus to which it is connected.

Convection Currents created by heating air, which then rises and pulls cooler air behind it. See **radiation**.

Conventional jack The type of hydraulic elevator mechanism whose cylinder must be installed in the ground. Installation must be perfectly vertical.

Conventional loan A mortgage loan not insured by a government agency (such as the Federal Housing Administration or Veterans Administration).

Convertibility The ability to change a loan from an adjustable rate schedule to a fixed rate schedule.

Cooking unit (counter mounted) A cooking appliance designed for mounting in or on a counter and consisting of one or more heating elements, internal wiring, and built-in or mountable controls.

Cooling load The amount of cooling required to keep a building at a specified temperature during the summer.

Cooling tower A piece of equipment, usually a large tower, that uses water to regulate air temperature in a facility by absorbing heat from air-conditioning systems and equipment and transferring the heat to the atmosphere.

Coped When the top and bottom flanges of the end(s) of a metal I-beam have been removed. This is done to permit it to fit within, and bolted to, the web of another I-beam in a "T" arrangement; should not

be done without the approval of a structural engineer.

Coped joint Cutting and fitting woodwork to an irregular surface.

Copper-clad aluminum conductors Conductors drawn from a copper-clad aluminum rod with the copper metallurgically bonded to an aluminum core. The copper forms a minimum of 10% of the cross-sectional area of a solid conductor or each strand of a stranded conductor.

Copper oxide A chemical that is created when copper is exposed to the atmosphere. Copper oxide is not an insulator and creates no hazard.

Corbel A triangular, decorative, and supporting member that holds a mantel or horizontal shelf.

Corbel block A block that projects a short distance from the face of a wall and provides support for other weight.

Core sample Test sample of concrete, usually 12 in. high and 6 in. in diameter. When properly cured, the core sample may be used to determine the strength of the concrete.

Corner bead A strip of formed sheet metal placed on outside corners of drywall before applying drywall "mud."

Corner block A decorative piece that is placed between the vertical side casing and the horizontal top casing; also see **base block**.

Corner boards Used as trim for the external corners of a house or other frame structure against which the ends of the siding are finished.

Corner braces Diagonal braces at the corners of the framed structure designed to stiffen and strengthen the wall.

Cornice The overhang of a pitched roof, usually consisting of a fascia board, a soffit, and appropriate trim moldings.

Corrugated roof panel A roofing sheet often made out of galvanized steel or fiberglass; shaped in alternating ridges and valleys.

Cosmetic damage Damage to an item or surface that affects the way an item looks but not the way it functions.

Cottage cheese texture See **popcorn texture**.

Counter balance The effect created by a force that is acting in opposition to another force; for example, the weight of an overhead door is counterbalanced by springs making it feel lighter when opened.

Counterflashings Counterflashings (or cap flashings), normally made of sheet metal, shield the exposed joints of base flashings. See **base flashings**.

Counterfort A foundation wall section (buttress) that strengthens (and is generally perpendicular to) a long section of foundation wall.

Course A row of shingles or roll roofing that runs the length of the roof; parallel layers of building materials such as bricks or siding laid horizontally.

Cove molding A molding with a concave face used as trim or to finish interior corners.

Cove piece (tile) Specialty tile trim piece used to trim corners in a tile surface. See **double bullnose**, and **tile base**.

Crawl space A shallow space below the living quarters of a house, normally enclosed by the foundation wall and having a dirt floor.

Cream In reference to concrete, cream is the thin layer of fine mixture that comes to the surface of concrete when the course aggregate is pushed down with a concrete finishing tool such as a gandy or bull float.

Credit rating A report ordered by a lender from a credit agency to determine a borrower's credit habits.

Cricket A second roof built on top of the primary roof to increase the slope of the roof or valley. A saddle-shaped, peaked construction connecting a sloping roof with a chimney.

Designed to encourage water drainage away from the chimney joint.

Crimp ring Used with fitting in PB pipe. The PB pipe is pressed firmly around the fitting with the crimp ring.

Cripple rafter A rafter that runs from a hip rafter to a valley rafter. A cripple rafter never reaches the wall top plate or the ridge board.

Cripples Short studs that are used to fill the gap under the windowsill and between the header and the top plate if there is a gap; also used in a non-bearing walls to fill the space above the door opening and the top plate when no header is required.

Critical visual tasks Visual tasks completed by building occupants, including reading and computer monitor use.

Cross bridging Diagonal bracing between adjacent floor joists, placed near the center of the joist span to prevent joists from twisting.

Cross tee A short metal T-beam used in suspended ceiling systems to bridge the spaces between the main beams.

Crown Most lumber is not chalkline straight but will bow slightly along its length. The upward bow is called the crown of the board. See **camber**.

Crowned stud When a wall is assembled on the floor, the framer places the crown of the stud upward. When the wall is stood in place, the convex or crown side of the stud is the "front," and the convex side of the stud is the "back."

Crowning Arranging all framing members so that all crowns are in the same direction. See **crowned stud**.

Crown molding A molding used on cornice or wherever an interior angle is to be covered, especially at the roof and wall corner.

Cubic foot (cu. ft.) A three-dimensional volume measurement equal to the amount contained by a cube that is 1-ft. wide, 1-ft. long, and 1-ft. high.

Cubic yard (cu. yd.) A three-dimensional volume measurement equal to the amount contained by a cube that is 1 yd. wide, 1 yd. long, and 1 yd. high. Each cu. yd. contains 27 cu. ft.

Culinary water Water that is fit for human consumption; also called potable water.

Cultured countertop A type of solid plastic material usually mixed with a pattern that imitates a type of stone. Culture marble countertops are probably the most common type, but cultured granite is also common. Cultured materials are also used to make tub and shower surrounds. See **cultured marble countertop, plastic laminate countertop, solid surface countertop, solid plastic countertop, stone countertop, tile countertop,** and **wood block countertop**.

Cultured marble countertop The most common type of cultured countertop contains swirls and color variations that imitate the look of a marble countertop. See **cultured marble countertop, plastic laminate countertop, solid plastic countertop, solid surface countertop, stone countertop, tile countertop**, and **wood block countertop**.

Cultured stone Masonry units made from man-made materials such as plaster or plastic. Cultured stone is shaped and colored to resemble natural stone but is much lighter weight.

Culvert A usually round, corrugated drain pipe.

Cupola A vent positioned at the ridgeline. Cupolas are often in the shape of a small house or dome, topped with a weathervane. Some cupolas are installed for decoration only and are nonfunctional.

Cupping A type of warping that causes boards to curl up at their edges.

Curb The short elevation of an exterior wall above the deck of a roof. Also, a two by six box (on the roof) on which a skylight is attached.

Curb stop Normally a cast iron pipe with a lid (at 5 in. in diameter) that is placed vertically into the ground, situated near the water tap in the yard, and where a water cutoff valve to the home is located (underground). A long pole with a special end is inserted into the curb stop to turn the water on or off.

Custom cabinets Milled cabinets that are built for a specific kitchen to match specified dimensions and design. See **built-in cabinets, mass-produced cabinets, and milled cabinets**.

Cut-in brace Nominal 2-in.-thick members, usually two-by-fours, cut in between each stud diagonally.

Cutout box An enclosure designed for surface mounting that has swinging doors or covers secured directly to and telescoping with the walls of the equipment.

Cut pile Carpet pile in which the ends are looped, both ends are attached to the carpet backing, and then the centers of the loops are cut. See **loop pile** and **pile**.

Dado A groove cut into a board or panel intended to receive the edge of a connecting board or panel.

Dalle glass A decorative composite glazing material made of individual pieces of glass that are embedded in concrete or epoxy.

Damper A metal "door" placed within the fireplace chimney; normally closed when the fireplace is not in use.

Damp proofing The black, tar-like, waterproofing material applied to the exterior of a foundation wall.

Daylight The end of a pipe (the terminal end) that is not attached to anything.

Daylight factor The ratio of exterior illumination to interior illumination, expressed as a percentage.

Daylighting The controlled admission of natural light into a space through glazing to reduce or eliminate electric lighting.

DC Electrical current (usually) provided by a battery.

Dead bolt An exterior security lock installed on exterior entry doors that can be activated only with a key or thumb-turn. Unlike a latch, which has a beveled tongue, dead bolts have square ends. The use of dead bolts is regulated by most fire codes.

Dead front (electrical) Not having live parts exposed to a person on the operating side of equipment.

Dead light The fixed, nonoperable window section of a window unit.

Dead load A permanent load consisting of all building parts and built-in fixtures that will be supported by a structural part.

Decimalized feet An expression of distance in ft. and decimal portions of ft. rather than ft. and in. For example, 6.25 ft. is the decimalized expression of 6 ft., 3 in.

Deck To install the plywood or wafer board sheeting on floor joists, rafters, or trusses.

Dedicated circuit A circuit that consists of a home run that connects to a single device.

Dedicated circuit An electrical circuit that serves only one appliance (i.e., dishwasher) or a series of electric heaters or smoke detectors.

Default Breach of a mortgage contract (not making the required payments).

De-humidistat A control mechanism used to operate a mechanical ventilation system based upon the relative humidity in the home.

Delamination The separation of the plies in a panel due to failure of the adhesive; usually caused by excessive moisture.

Demand factor The ratio of the maximum demand of a system, or part of a system, to the total connected load of a system or to the part of the system under consideration.

Densely occupied space An area that has a design occupant density of 25 people or more per 1000 sq. ft. (40 sq. ft. or less per person). Most codes specify ventilation requirements for these spaces.

Designer A person who recommends aesthetically pleasing combinations of shapes and shades for the interior or the exterior of a building.

Design light output The light output of lamps at 40% of their useful life.

Detail One of the five basic views found on a plan. A detail is a close-up (i.e., large scale) view of some part of a section view, used to show exactly how parts connect together.

Development footprint The area affected by project activity. Hardscape, access roads, parking lots, nonbuilding facilities, and the building itself are all included in the development footprint.

Device (electrical) A unit of an electrical system that is intended to carry but not utilize electric energy.

Diagonal bracing A reinforcing member that is attached at an angle to provide lateral strength to another member such as a rafter.

Diagonal wood floor installation An installation where wood strips are installed in a pattern that runs diagonally to at least one of the walls. See **herringbone wood floor installation** and **straight wood floor installation**.

Diagramming The process of drawing a floor plan and sometimes elevations that include dimensions and other important information such as the location of doors, windows, outlets, switches, and so on.

Diameter A straight line segment that passes through the center of a circle and terminates at the outer edges of the circle. It is the longest line segment that will fit inside a circle. The diameter is equal to twice the length of the radius.

Dimensional beam Made from dimensional lumber usually available in rough cut or surfaced finishes.

Dimensions Dimensions are the measurements of a room or object. From plans, dimensions can be determined using the measurement on the drawing and the scale of the drawing or by reading the dimensions printed on the drawing.

Direct costs Construction costs that can be specifically identified with a project or with a unit of production within a project.

Disconnect A large (generally 20 amp) electrical on-off switch.

Disconnecting means A device, group of devices, or other means by which the conductors of a circuit can be disconnected from their source of supply.

Discount rate A mortgage interest rate that is lower than the current rate for a certain period of time (e.g., 2.00% below variable rate for 2 years).

Door gasket Insulating material placed around the door jamb where the slab meets the jamb or doorstop when closed. The door gasket is used on exterior doors.

Door hardware The latch, door knob, and striker plate.

Doorjamb (interior) The surrounding case into which and out of which a door closes and opens. It consists of two upright pieces, called side jambs, and a horizontal head jamb. These three jambs have the doorstop installed on them.

Door operator An automatic garage door opener.

Doorstop The wooden style that the door slab will rest upon when it's in a closed position.

Dormer An opening in a sloping roof, the framing of which projects out to form a vertical wall suitable for windows or other openings.

Double bullnose (tile) A specialty trim piece with rounded corners on both sides. See **cove piece** and **tile base**.

Double glass Window or door in which two panes of glass are used with a sealed air space between; also known as insulating glass.

Double-hung window A window with two vertically sliding sashes, both of which can move up and down.

Double pull A method for using the ladder to climb a lower roof section, then pulling the ladder onto the roof for use when climbing a second-higher roof section.

Dowel A stick of wood or metal that fits into corresponding holes to attach two pieces of material together.

Downflow furnace A furnace that forces air down and out the bottom of it.

Down payment The difference between the sales price and the mortgage amount. A down payment is usually paid at closing.

Downspout A pipe, usually of metal, for carrying rainwater down from the roof's horizontal gutters.

Drafter A person who translates the ideas provided to them by an architect or designer into accurate plans that may be used for construction. In this age of computers, fine drafting has become a lost art.

Drain field A series of pipes through which waste is released into the soil after it has been treated in the septic tank. The size of the drain field depends on the number of people being serviced by the system and the ability of the soil to absorb liquid.

Drain tile A perforated, corrugated plastic pipe laid at the bottom of the foundation wall and used to drain excess water away from the foundation. It prevents ground water from seeping through the foundation wall; sometimes called perimeter drain.

Draw The amount of progress billings on a contract that is currently available to a contractor under a contract with a fixed payment schedule.

Drip A member of a cornice or other horizontal exterior finish course that has a projection beyond the other parts for throwing off water.

Drip A groove in the underside of a sill or drip cap to cause water to drop off on the outer edge instead of drawing back and running down the face of the building.

Drip cap A molding or metal flashing placed on the exterior topside of a door or window frame to cause water to drip beyond the outside of the frame; also known as drip edge.

Drip irrigation A high-efficiency method in which water is distributed at low pressure through buried mains and submains. From the submains, water is distributed to the soil from a network of perforated tubes or emitters. Drip irrigation is a type of microirrigation.

Drip loop A loop that is made in the drop wires just in front of the weather head. The drip loop prevents water from dripping down the wires and into the weather head.

Drop siding Shiplap siding with special shaping on its face; for example, it may have a rounded face to look like logs.

Drop wire A wire that connects the transformer to the structure. The drop wire for a residential structure usually contains two current carrying leads and a neutral.

Dry in To install the black roofing felt (tar paper) on the roof.

Drying Process used to remove moisture from lumber. The moisture content in the lumber is reduced to the average amount it will maintain when used in building construction. After drying, many lumber mills seal the lumber with stain or wax to prevent them from absorbing moisture again. Surface drying and kiln drying are two methods used for drying lumber.

Dry looper A machine used in the manufacture of composition shingles that is designed to allow air to circulate around roofing material to dry the material.

Dry pipe valve A valve that automatically controls the water supply to a sprinkler system so that the system beyond the valve is normally maintained dry.

Drywall (or gypsum wallboard [GWB], sheet rock or plasterboard) A manufactured panel made out of gypsum plaster and encased in a thin cardboard; usually ½-in. thick and 4 ft. × 8 ft. or 4 ft. × 12 ft. in size. The panels are nailed or screwed onto the framing and the joints are taped and covered with a "joint compound." "Green board"–type drywall has a greater resistance to moisture than regular (white) plasterboard and is used in bathrooms and other "wet areas."

Ducts A tunnel made of galvanized metal or rigid fiberglass that carries air from the heater or ventilation opening to the rooms in a building.

Ducts (ventilation) The heating system; usually round or rectangular metal pipes installed for distributing warm (or cold) air from the furnace to rooms in the home.

Due-on-sale A clause in a mortgage contract requiring the borrower to pay the entire outstanding balance upon sale or transfer of the property.

Dumbwaiter A hoisting and lowering mechanism equipped with a car that moves in guides in a substantially vertical direction. Dumbwaiters are used exclusively for carrying materials, not people.

Duplex An electrical outlet with two ports.

Durable goods Goods with a useful life of 2 years or more and are replaced infrequently or may require capital program outlays. Examples include furniture, office equipment, appliances, external power adapters, televisions, and audiovisual equipment.

Durable goods waste stream Durable goods leaving the project building, site, and organization that have fully depreciated and have reached the end of their useful lives for normal business operations.

Dura board, dura rock A panel made out of concrete and fiberglass usually used as a ceramic tile backing material; commonly used on bathtub decks; sometimes called wonderboard.

Dust-tight (electrical) Constructed so that dust will not enter the enclosing case under specified test conditions.

Dutch door A door that contains two half slabs mounted one above the other. Each slab is attached to the jamb with hinges and may swing independently of the other.

Duty (continuous) Operation at a substantially constant load for a indefinitely long time.

Duty (intermittent) Operation for alternate intervals of (1) load and no load; or (2) load and rest; or (3) load, no load, and rest.

Duty (periodic) Intermittent operation in which the load conditions are regularly recurrent.

Duty (short-time) Operation at a substantially constant load for a short and definite, specified time.

Duty (varying) Operation at loads and for intervals of time, both of which may be subject to wide variations.

Dwelling Any building occupied in whole or in part as the temporary or permanent home or residence of one or more families.

Dwelling unit One or more rooms in a dwelling that are arranged, designed, used, or intended for use by one or more families.

DWV (drain-waste-vent) The section of a plumbing system that carries water and sewer gases out of a home.

Earnest money A sum paid to the seller to show that a potential purchaser is serious about buying.

Earthquake strap A metal strap used to secure gas hot water heaters to the framing or foundation of a house; intended to reduce the chances of having the water heater fall over in an earthquake and causing a gas leak.

Easement A formal contract that allows a party to use another party's property for a specific purpose. (e.g., a sewer easement might allow one party to run a sewer line through a neighbor's property).

Eave The horizontal exterior roof overhang.

Ecological restoration The process of assisting in the recovery and management of ecological integrity, including biodiversity, ecological processes and structures, regional and historical content, and sustainable cultural practices.

Edge grain shingles Shingles made from lumber whose annual rings form at least a 45-degree angle at the face.

Egress A means of exiting the home. Most codes require an egress window in every bedroom and basement. Normally a 4 ft. × 4 ft. window is the minimum size required.

Elastomeric roof system Stretchable single membrane roof system made from either plastic PVC or rubber

EPDM (ethylene propylene diene monomer).

Elbow (ell) A plumbing or electrical fitting that lets you change directions in runs of pipe or conduit.

Electrical circuit A group of outlets or other electrical devices that connect to a fuse box or breaker panel through a common lead.

Electrical entrance package The entry point of the electrical power including (a) the "strike" or location where the overhead or underground electrical lines connect to the house, (b) the meter that measures how much power is used, and (c) the "panel" or "circuit breaker box" (or "fuse box") where the power can be shut off and where overload devices, such a fuses or circuit breakers, are located.

Electrical rough Work performed by the electrical contractor in conjunction with the plumber and heating contractor or after they have completed their work. Normally all electrical wires and outlet, switch, and fixture boxes are installed before insulation is put in.

Electrical trim Work performed by the electrical contractor when the house is nearing completion. The electrician installs all plugs, switches, light fixtures, smoke detectors, appliance "pig tails," bath ventilation fans and also wires the furnace, and "makes up" the electric house panel. The electrician does all work necessary to get the home ready for and to pass the municipal electrical final inspection.

Electric lateral The trench or area in the yard where the electric service line (from a transformer or pedestal) is located, or the work of installing the electric service to a home.

Electric resistance coils Metal wires that heat up when an electric current passes through them and are used in baseboard heaters and electric water heaters.

Electrostatic filter A filter that magnetically attracts dust to its surface.

Elemental mercury Pure mercury (rather than a mercury-containing compound), the vapor of which is commonly used in fluorescent and other lamp types.

Elevation One of the five basic views found on a plan. Elevation is an eye-level view of a surface on the building.

Elevation sheet The page on the blueprints that depicts the house or room as if a vertical plane were passed through the structure.

Elevator A hoisting and lowering mechanism equipped with a car or platform that moves in guides in a substantially vertical direction and that serves two or more floors of a building.

Emergency board up The process of securing a structure from weather and unwanted entry. Damaged doors and windows or other easily accessible openings are typically covered with plywood and openings in the roof are typically covered with plastic. This work is usually done under contract with a city or county agency.

Emergency interlock release switch (elevator) A device to make the door, gate electric contacts, or door interlocks inoperable in case of emergency.

Emissivity Emissivity is the ratio of the radiation emitted by a surface to the radiation emitted by a black body at the same temperature.

Enclosed (electrical) Surrounded by a case, housing, fence, or wall that prevents persons from accidentally contacting energized parts.

End-matched board End-matched boards have tongues and grooves along their sides and the ends.

Energized (electrical) Electrically connected to a source of voltage.

Energy audit An energy audit identifies how much energy is used in a building for what purposes and identifies opportunities for improving efficiency and reducing costs.

ENERGY STAR rating A measure of a building's energy performance compared with buildings with similar characteristics, as determined by use of the ENERGY STAR® Portfolio Manager.

Engineer A person licensed to practice the profession of engineering under state law.

Environmental tobacco smoke (ETS), or secondhand smoke Airborne particles that are emitted both directly from cigarettes, pipes, and cigars and indirectly exhaled by smokers.

EPDM (ethylene propylene diene monomer) A synthetic rubber material used to make elastomeric roof membranes.

Epoxy A two-part resin that, when mixed, forms a tight cross-linked polymer. Epoxy forms a hard, tough surface that is highly resistant to corrosion.

Epoxy finish A concrete finish made by spreading an epoxy adhesive over cured concrete then placing aggregate over the epoxy, which glues the aggregate to the underlying concrete.

Epoxy injection technique The process of injecting epoxy resin into a crack. A good epoxy joint is usually stronger than the material it replaces.

Equity The "valuation" that you own in your home (i.e., the property value less the mortgage loan outstanding).

Erosion The process by which the materials of Earth's surface are loosened, dissolved, or worn away and transported by natural agents.

Escrow The handling of funds or documents by a third party on behalf of the buyer or seller.

Escutcheon An ornamental plate that fits around a pipe extending through a wall or floor to hide The cut-out hole.

Estimate The amount of labor, materials, and other costs that a contractor anticipates for a project as summarized in the contractor's bid proposal for the project.

Estimating The process of calculating the cost of a project. This can be a formal and exact process or a quick and imprecise process.

Etched glass Hardened glass that has been sandblasted or otherwise engraved to form a pattern or design in the glass.

European-style cabinet door hinge A cabinet hinge that is mortised into the back of the door on one side and attached to the cabinet box on the other side. Once installed, it can be adjusted in a variety of directions. The European-style cabinet door hinge was designed for use on the frameless style cabinets that were developed in Europe after World War II, but it is also found on some higher-quality framed-style cabinets; also called a six-way adjustable hinge and a recessed hinge. See **standard cabinet door hinge**.

European wallpaper roll A roll of wallpaper that contains about 28 sq. ft. of material.

Evaporative cooler System that cools by drawing air through moist filters that transfer moisture into the air. Evaporative coolers are effective only in regions with relatively low humidity.

Evaporator coil The part of a cooling system that absorbs heat from air in your home. See **condensing unit**.

Evapotranspiration Water lost through transpiration through plants plus water evaporated from the soil.

Evapotranspiration rate (ET) The amount of water lost from a vegetated surface in units of water depth. It is expressed in millimeters per unit of time.

Excavator A contractor or piece of equipment that moves the soil out of the area where footings, foundations, or utility lines will be placed and backfills the soil around these parts once they are in place.

Expanded foam A foam that contains small beads with air voids around the beads. When used as insulation, expanded foam should only be installed above grade.

Expansion joint Material installed to permit movement between structural members.

Expansive soils Earth that swells and contracts depending on the amount of water that is present ("Betonite" is an expansive soil).

Exposed aggregate finish A method of finishing concrete that washes the cement or sand mixture off the top layer of the aggregate, usually gravel.

Exposed (as applied to live electrical parts) Capable of being inadvertently touched or approached at nearer than a safe distance by a person. It is applied to parts that are not suitably guarded, isolated, or insulated.

Exposed (as applied to wiring methods) On or attached to the surface or behind panels designed to allow access.

Extras Additional work requested of a contractor, not included in the original plan, that will be billed separately.

Extruded foam Foam that is smooth, with no beads or voids. Because extruded foam will not absorb water,

it can be installed as insulation above or below grade.

Face (cabinet) The cabinet face frame, doors, and drawer fronts. See **box** and **face frame**.

Faced concrete To finish the front and all vertical sides of a concrete porch, step(s), or patio. Normally the "face" is broom finished.

Face frame A frame placed over the cabinet box against which the doors and drawers will rest. The face frame is made from vertical stiles and horizontal rails. See **box** and **face**.

Face nailing To install nails into the vertical face of a bearing header or beam. See **blind nailing**.

Facing brick The brick used and exposed on the outside of a wall.

Factory-built fireplace A prebuilt firebox assembly that includes a heat exchanger, air movement equipment, and the flue assembly; also called a zero clearance fireplace.

Fairtrade A product certification system overseen by FLO International. (FLO-CERT GmbH is an independent international certification company. FLO assists in the socioeconomic development of producers in the global south.) FLO allows people to identify products that meet agreed upon environmental, labor, and development standards.

False tread and riser A nonstructural, decorative tread and riser assembly that is placed over the structural tread and riser when carpet runs down the center of a stair. False tread and risers are typically stain-grade wood that gives the appearance of higher-grade wood treads and risers at a lower cost. See **balustrade**.

Fascia Horizontal boards attached to rafter or truss ends at the eaves and along gables. Roof drain gutters are attached to the fascia.

Feeder Connects the meter base to the breaker panel(s) or fuse box(es); usually contains four cables that are twisted together.

Felt Tar paper that is installed under the roof shingles.

Ferrule Metal tubes used to keep roof gutters "open." Long nails (ferrule spikes) are driven through these tubes and hold the gutters in place along the fascia of the home.

Festoon lighting A string of outdoor lights that is suspended between two points.

FHA strap Metal straps that are used to repair a bearing wall "cutout" and to "tie together" wall corners, splices, and bearing headers. Also, they are used to hang stairs and landings to bearing headers.

Fiber and cement shingle A shingle made from a combination of wood fiber and Portland cement. Fiber and cement shingles can be made to resemble slate, tile, or wood.

Fiberglass Glass filaments that are formed by pulling or spinning molten glass into random lengths.

Field measure To take measurements (cabinets, countertops, stairs, shower doors, etc.) on-site instead of using the blueprints.

Field total The amount of material, including waste that will be required in the field in order to complete the job.

Fillet A small piece of decorative wood that fills the space between balusters in a bottom rail. See **balustrade**.

Final inspection A review by the building inspector after the interior and exterior construction is complete to check for any problems that may endanger the health or safety of the building occupants. In most areas, a certificate indicating that the final inspector has been successfully completed is required before the project can be considered completed.

Finger joint A manufacturing process of interlocking two shorter pieces of wood end to end to create a longer piece of dimensional lumber or molding; often used in jambs and casings and are normally painted (instead of stained).

Finish May refer to the plumbing, electrical, carpentry, or HVAC work that is visible when construction is complete.

Finish coat Final coat of any material on a surface. The third coat of common stucco is the finish coat and contains the texture and may contain the pigment. If the finish coat does not contain pigment, the surface of the stucco must be painted when dry. For stucco, see also **brown coat** and **scratch coat**.

Finish electrical Any electrical part that will be installed after the walls and ceiling are finished.

Finish nail A small-headed nail that can be recessed below the surface of the wood, leaving only a small hole that can be puttied easily.

Fire block Short horizontal members sometimes nailed between studs, usually about halfway up a wall. See also **fire stop**.

Fire box The interior of a fireplace system built of heat-resistant materials which contains the fire and radiates heat in the room. Can be made from a variety of materials including special fire brick, prefabricated masonry panels, or metal.

Fire brick Brick made of refractory ceramic material that will resist high temperatures. Used in a fireplace and boiler.

Fired brick Masonry unit made from clay that is formed and then baked at a high enough temperature to cause a partial melting or glazing on the surface. This glaze provides a seal that protects the brick from moisture.

Firefighter telephone jack A special phone jack that can be built into an elevator car operating panel or hall station that enables communication with firefighters.

Fireplace chase flashing pan A large sheet of metal that is installed around and perpendicular to the fireplace flue pipe. Its purpose is to confine and limit the spread of fire and smoke to a small area.

Fire-resistive or fire-rated Applies to materials that are not combustible in the temperatures of ordinary fires and will withstand such fires for a rated period. Building codes establish required fire ratings.

Fire-retardant chemical A chemical or preparation of chemicals used to reduce the flammability of a material or to retard the spread of flame.

Fire section An area with sprinklers within a building that is separated from other areas by noncombustible construction having at least a 2-hour fire resistance rating.

Fire stop A solid, tight closure of a concealed space, placed to prevent the spread of fire and smoke through such a space. In a frame wall, this will usually consist of two-by-fours cross-blocking between studs. Work performed to slow the spread of fire and smoke in the walls and ceiling (behind the drywall) includes stuffing wire holes in the top and bottom plates with insulation and installing blocks of wood between the wall studs at the drop soffit line. This is integral to passing a rough frame inspection. See **fire block**.

Fire tape The process in which drywall is finished to provide fire protection only and not to provide a smooth finish wall. Fire-taped drywall has tape embedded along all joints that are then covered with one additional layer of mud. Fasteners are also covered with mud.

Fire wall A wall that has been designed to resist the spread of fire. Fire walls in homes are typically required between the garage and living space. Fire walls are usually rated by the hours they are designed to resist the spread of fire.

Fishplate (gusset) A wood or plywood piece used to fasten the ends of two members together at a butt joint with nails or bolts. Sometimes used at the junction of opposite rafters near the ridgeline; sometimes called a gang nail plate.

Fish tape A long strip of spring steel used for fishing cables and for pulling wires through a conduit.

Fissuring A method of imparting a set of ragged depressions into the face of acoustical tile or panels during manufacture for appearance and acoustical performance.

Five-quarter A term used for exterior decking material. Refers to the one and one-quarter or "five-quarter" thickness of the material.

Fixed price contract A contract with a set price for the work. See **time and materials contract**.

Fixed rate A loan where the initial payments are based on a certain interest rate for a stated period. The rate payable will not change during this period regardless of changes in the lender's standard variable rate.

Fixed-rate mortgage A mortgage with an interest rate that remains the same over the years.

Fixed window A window unit that does not open; also may be referred to as a picture window.

Flagstone (flagging or flags) Flat stones (1 to 4 in.thick) used for walks, steps, floors, and vertical veneer (in lieu of brick).

Flakeboard A manufactured wood panel made out of 1- to 2-in. wood chips and glue. Often used as a substitute for plywood in the exterior wall and roof sheathing; also called OSB or wafer board.

Flame retention burner An oil burner designed to hold the flame near the nozzle surface.

Flanges Parallel edges that are perpendicular to the center web of a steel beam (girder).

Flashing Any piece of material installed to prevent water from penetrating into the structure around doors, windows, chimneys, and roof edges.

Flashing A flaw in drywall that occurs when reflected light reveals a difference between areas where drywall mud was applied over tape or fasteners and areas where mud was not applied.

Flat grain shingles Shingles made from lumber whose annual rings form less than a 45-degree angle at the face.

Flat-laid countertop Type of plastic laminate countertop with no integral backsplash, a flat smooth surface, and usually a square front edge. See **plastic laminate countertop** and **post-formed countertop**.

Flat mold Thin wood strips installed over the butt seam of cabinet skins.

Flat paint An interior paint that contains a high proportion of pigment and dries to a flat or lusterless finish.

Flat panel A flat, thin plywood panel used in a frame and panel cabinet door. See **frame and panel cabinet panel**, **raised panel**, and **slab cabinet door**.

Flat roof A roof style that appears flat but actually has a slight slope to allow drainage of precipitation. As a minimum, roofs should be sloped ¼ in. per ft. Do not forget to consider deflection due to loading when designing a roof.

Flat tiles A tile shingle with a flat surface. The surfaces of flat tiles often have a grain simulation and the sides are usually rabbeted and grooved.

Flat truss A roof or floor truss with horizontal top and bottom chords reinforced with diagonal members between them.

Flatwork A common word for concrete floors, driveways, basements, and sidewalks.

Flitch A large piece of lumber cut out of a log that is then sawn into boards or veneer strips.

Float Hand tool used to provide an even texture to concrete or plaster surfaces before they set.

Floating The next-to-last stage in concrete work, when you smooth off the job and bring water to the surface by using a hand float or bull float.

Floating wall A nonbearing wall built on a concrete floor. It is constructed so that the bottom two horizontal plates can compress or pull apart if the concrete floor moves up or down; normally built on basements and garage slabs.

Flood coat Heavy, smooth asphalt coating mopped over the cap sheet of a multiple-ply membrane roof to provide a smooth surface. The flood coat must be protected from sun damage by painting it with a UV coat or by covering it with aggregate.

Flood plane structure A structure that sits atop columns that raise the main floor above the flood plane.

Floor plan One of the five basic views found on a plan; also known as plan view. A floor plan is a view as though you are looking directly down on a building with the top removed so you can see the layout of the floor including walls and fixtures.

Floor system Includes the framing support members such as floor joists or floor trusses and the sheathing that provides the floor system surface. This is a structural part.

Floor truss May be used instead of regular joists or I-joists; floor trusses are generally placed on wider centers and are deeper and more expensive than other joists. Floor trusses are designed to allow plumbing,

electrical, and heating runs to be placed inside of them instead of below them, as is often required in other joists.

Flue A large pipe through which fumes escape from a gas water heater, furnace, or fireplace. Normally these flue pipes are double-walled, galvanized sheet metal pipe and sometimes referred to as a B-vent. Fireplace flue pipes are normally triple walled. In addition, nothing combustible shall be within 1 in. from the flue pipe.

Flue cap A cap placed on the top opening of the flue in such a way as to permit proper ventilation of the inner chambers of the flue pipe and at the same time prevent moisture or small animals from entering the flue.

Flue collar See **flue cap**.

Flue damper An automatic door located in the flue that closes it off when the burner turns off; its purpose is to reduce heat loss up the flue from the still-warm furnace or boiler.

Flue lining Two-ft. lengths of fire clay or terra-cotta pipe (round or square) and usually made in all ordinary flue sizes; used for the inner lining of chimneys with the brick or masonry work done around the outside. Flue linings in chimneys run from 1 ft. below the flue connection to the top of the chimney.

Fluorescent lighting A fluorescent lamp is a gas-filled glass tube with a phosphor coating on the inside. Gas inside the tube is ionized by electricity that causes the phosphor coating to glow. Lamps normally have two pins that extending from each end.

Flush door slab Door with a flat, smooth face with no panels or decoration. Flush doors may have a solid or hollow core.

Flush tile Flat ceiling tiles with neither embossed designs nor recessed edges.

Fluted casing A casing that contains a series of round reliefs called flutes along its length. Fluted casings are designed to look like fluted columns.

Fly rafters A gable rafter that is located under the overhang part of the roof sheathing on the gable end. They are not directly supported by the exterior wall; also sometimes referred to as the barge rafter or barge board.

Foam foundation sill plate sealer Thin foam strip placed on top of the foundation and under the foundation sill plate. This strip is actually pushed down over the bolts and fills in any gaps between the foundation and the foundation sill plate.

Footer (footing) A continuous concrete pad that is installed before and supports the foundation wall or monopost.

Footing The base upon which the structure will stand; it rests on the soil. A footing ultimately supports all of the weight of the structure; it is a structural part.

Forced air system A common form of heating with natural gas, propane, oil, or electricity as a fuel. Air is heated in the furnace and distributed through a set of metal ducts to various areas of the house.

Form A temporary structure erected to contain concrete during placing and initial hardening.

Formica A brand name of a common type of plastic laminate material. The term Formica is often used in the industry when referring to plastic laminate. See **plastic laminate countertop**.

Foundation The supporting portion of a structure below the first floor construction, or below grade, including the footings.

Foundation sill plate A piece of lumber (some codes require treated lumber) that is used between the foundation and the framing. It is attached to the foundation with anchor bolts.

Foundation ties Metal wires that hold the foundation wall panels and rebar in place during the concrete pour.

Foundation waterproofing A high-quality, below-grade moisture protection that is used for below-grade exterior concrete and masonry wall damp-proofing to seal out moisture and prevent corrosion.

Four-way inspection An inspection of the rough-in of four trades including framing, plumbing, HVAC, and electric. This inspection must be completed before the walls or ceilings are closed in.

Fractured composition shingles A composition shingle that has been torn by the impact from a hailstone. The fractures often radiate out from the center of the hailstone impact in a spider-web pattern. Also see **bruised composition shingles** and **granular loss**.

Frame and panel cabinet door A cabinet door that consists of a frame that surrounds a panel. The panel may be glass, a veneered plywood flat panel, or a solid wood raised panel. See **slab cabinet door**.

Framed-style cabinets A cabinet style in which a face frame is attached to the cabinet box. See **frameless-style cabinets**.

Framed to square A term to denote that the building framing has been completed to the point that it is ready for the roof frame to be built.

Frame inspection The act of inspecting the home's structural integrity and it's compliance with local municipal codes.

Frameless style cabinet A cabinet style that has no face frame attached to the cabinet box; often called a European-style cabinet because it was developed in Europe during the reconstruction following World War II as an alternative to the more labor-intensive framed-style cabinet. See **framed-style cabinets**.

Framer The carpenter contractor that installs the lumber and erects the frame, flooring system, interior walls, backing, trusses, rafters, and decking and installs all beams, stairs, soffits, and work related to the wood structure of the home. The framer builds the home according to the blueprints and must comply with local building codes and regulations.

Framing Lumber used for the structural members of a building, such as studs, joists, and rafters.

Framing tie Members that connect the bottoms of opposing rafters together to prevent them from moving outward. Ceiling joists are commonly used as framing ties.

Freezeless hose bibb A faucet designed to supply water to the outside of the structure without danger of freezing in cold temperatures. The faucet is located on the outside of the structure but the valve portion is located inside the heated structure.

Frieze board In house construction, a horizontal member connecting the top of the siding with the soffit of the cornice.

Frost lid A round metal lid that is installed on a water meter pit.

Frost line The depth of frost penetration in soil or the depth at which the earth will freeze and swell. This depth varies in different parts of the country.

Full backsplash A backsplash that runs from the countertop to the bottom of the upper cabinet. See **backsplash** and **block backsplash**.

Full basement structure A structure that has a basement level—the floor usually positioned below ground level, beneath the main level.

Full cutoff Describes a luminaire having a light distribution in which the candela per 1000 lamp lumens does not numerically exceed 25 (2.5%) at or above an angle of 90 degrees above nadir, and 100 (10%) at or above a vertical angle of 80 degrees above nadir. Light pollution is becoming a major issue in urban areas and one that you may wish to familiarize yourself with.

Full disclosure For products that are not formulated with listed suspect carcinogens, full disclosure has two components: (1) disclosure of all ingredients (both hazardous and nonhazardous) that make up 1% or more of the undiluted product and (2) use of concentration ranges for each of the disclosed ingredients. Full disclosure for products that are formulated with listed suspect carcinogens has three components: (1) disclosure of listed suspect carcinogens that make up 0.1% or more of the undiluted product, (2) disclosure of all remaining ingredients (both hazardous and nonhazardous) that make up 1% or more of the undiluted product, and (3) use of concentration ranges for each of the disclosed ingredients. Previously, full disclosure was not a concern to estimators, but nowadays, both clients and workers are requiring that this information be available on all jobsites.

Full extension glide Hardware attached between a cabinet drawer and the cabinet box that allows the drawer to be pulled completely out of the cabinet box. See **glide**.

Full height cabinet Any cabinet that runs the full height from the floor to the level of the upper unit. See **lower unit**, **upper unit**, and **vanity cabinet**.

Full-time equivalent (FTE) Generally, a 40-hours-per-week job. Part-time and overtime positions have FTE values based on their hours per week divided by 40. Multiple shifts are included or excluded depending on the intent and requirements of the credit. Estimators frequently develop labor costs based on FTEs, but beware as fringe benefits are often different for part-time and overtime hours.

Fully shielded An exterior light fixture that is shielded or constructed so that its light rays project below the horizontal plane passing through the lowest point on the fixture from which light is emitted.

Furniture, fixtures, and equipment (FFE) Includes all items that are not base building elements, such as lamps, computers and electronics, desks chairs, and tables. If your contract requires that you, the contractor, must provide FFE, make sure there is a full understanding with the owner as to exactly what this means. Also make sure all purchase orders make work items, such as delivery, clear (e.g., sidewalk or installed and time of delivery, such as normal work hours or at night so as not to interfere with construction).

Furring strips Strips of wood, often 1 in. by 2 in., used to shim out and provide a level-fastening surface for a wall or ceiling.

Fuse A device often found in older homes designed to prevent overloads in electrical lines. This protects against fire. See also **circuit breakers**.

Gable The end, upper, triangular area of a home, beneath the roof.

Gable end truss A truss used at the ends of a gable roof. It has vertical members that are spaced to allow convenient attachment of the exterior wall sheathing.

Gable roof A roof style consisting of two sides that slope in opposite directions down from the peak or ridge. The roof end forms an inverted V and is filled in with a triangular-shaped gable end wall.

Gable vent Vents placed in the gable ends of the roof. Gable vents facilitate the flow of air in the attic while protecting it from insects and the weather.

Galvanized A generic term used to describe steel coated with zinc applied in an electrogalvanizing or dipping process.

Gambrel roof A roof style consisting of two sides that meet at the ridge and slope in opposite directions. Each side has two sections, the lower section having a steeper slope than the upper section. The Gambrel roof is often used on barns.

Gambrel truss A truss used to make a Gambrel roof. It functions in the same way as a gable truss.

Gandy A tool that uses a screen to press the course aggregate downward while leaving the fine aggregate at the surface. A flat gandy is dropped lightly over the entire surface. A rolling gandy uses a screen shaped like a barrel and is rolled across the entire surface. Widely used in residential construction, its use is discouraged by many structural engineers because it can severely damage the concrete unless used skillfully.

Gang More than one switch installed in a single electrical box.

Gang nail plate A steel plate attached to both sides at each joint of a truss; sometimes called a **fishplate** or **gussett**.

Gasket Soft, pliable material used to prevent joint leakage.

Gas lateral The trench or area in the yard where the gas line service is located, or the work of installing the gas service to a home.

Gas piping systems The gas service piping, meter piping, and distribution piping.

Gate valve A valve that lets you completely stop but not modulate the flow within a pipe.

Gauge Refers to the thickness of metal or wire. A heavier gauge means that the metal is thicker, but it is denoted by a smaller number. The following is an example:

Gauge	Std. steel thickness (in.)	Aluminum thickness (in.)
3	0.2391	0.2294
5	0.2092	0.1819
7	0.1793	0.1443
15	0.0673	0.0571

General contractor A contractor who enters into a contract with the owner of a project for the construction of the project and who takes full responsibility for its completion. The contractor may enter into subcontracts with others for the performance of specific parts or phases of the project.

Geotechnical engineer See **soils engineer**.

GFCI, or GFI (ground fault circuit interrupter) An ultrasensitive plug designed to shut off all electrical currents. Used in bathrooms, kitchens, exterior waterproof outlets, garage outlets, and "wet areas." Has a small reset button on the plug.

Girder A large or principal beam of wood or steel used to support loads along its length.

Glare Any excessively bright source of light within the visual field that creates discomfort or loss in visibility.

Glazed tile A tile shingle with a color glaze compound put on its surface that produces a smooth and shiny face. Glazed clay tiles are baked. Glazed concrete tiles dry chemically. Glaze usually adds significantly to the cost of clay tiles but adds only moderately or not at all to the cost of concrete tiles.

Glazed wall tile Tile that comes in a variety of sizes but are usually about 4 in. × 4 in. and typically have a high-gloss or matte glaze applied to the finish surface. See **ceramic mosaic tile** and **quarry tile**.

Glazing The process of installing glass, which is commonly secured with glazier's points and glazing compound.

Glide Hardware attached between a cabinet drawer and the cabinet box that holds the drawer in a level position as the drawer is pulled out of and pushed into the cabinet box. See **full extension glide**.

Globe valve A valve that lets you adjust the flow of water to any rate between fully on and fully off. Also see **gate valve**.

Gloss enamel A finishing paint material; forms a hard coating with maximum smoothness of surface and dries to a sheen or luster (gloss).

Glued laminated beam (Glulam) A structural beam composed of wood laminations or lams. The lams are pressure bonded with adhesives to attain a typical thickness of 1½ in.

Glue-up tile A roof tile that is glued into place. Glue-up tile is usually 12 in. × 12 in.

Goose neck A long, curving handrail piece that is used to step down and make a long vertical transition between handrail parts on a stair balustrade. See **balustrade**, **one-quarter turn**, **top rail**, and **volute**.

Grade The elevation at ground level or at any given point.

Grade The work of leveling dirt, as in "to grade."

Grade The designated quality of a manufactured piece of wood.

Grade beam A foundation wall that is poured at level with or just below the grade of the earth. An example is the area where the 8- or 16-ft. overhead garage door "blockout" is located or a lower (walkout basement) foundation wall is poured.

Grade stake A stake that is placed in the ground and marked at the point where the grade should be found once the building part is in place. Grade stakes are often placed and marked to indicate where the top of the concrete will be located once the pour is complete. Grade stakes used as a guide for establishing the final level of the concrete are usually pulled out and their holes filled with wet concrete once the wet concrete has been leveled at the proper grade.

Graduated payment mortgage (GPM) A fixed-rate, fixed-schedule loan. It starts with lower payments than a level payment loan; payments rise annually, with the entire increase being used to reduce the outstanding balance. The increase in payments may enable the borrower to pay off a 30-year loan in 15 to 20 years or less.

Grain The direction, size, arrangement, appearance, or quality of the fibers in wood.

Granular loss Granular loss occurs when mineral granules embedded in a composition shingle are loosened by the impact from a hailstone when the hailstone does not bruise or fracture the shingle. See **bruised composition shingles** and **fractured composition shingles**.

Grass cloth wallpaper Wall covering made of loosely woven vegetable fibers.

Gray water, or gray water Defined by the Unified Plumbing Code (UPC) as "untreated household wastewater that has not come into contact with toilet waste. Gray water includes used water from bathtubs, showers, bathroom wash basins, and water from clothes-washer and laundry tubs. It shall not include wastewater from kitchen sinks and dishwashers." Other agencies define gray water differently (e.g., some municipalities include kitchen sink and dishwasher wastewater in the definition of gray water).

Green cleaning The use of cleaning products and practices that have less adverse environmental effects than conventional products and practices. Green cleaning may add cost to your project.

Greenfield conduit A type of flexible metal conduit.

Grid The completed assembly of main and cross tees in a suspended ceiling system before the ceiling panels are installed; also the decorative slats (muntin) installed between glass panels.

Grommet A circular eyelet that reinforces a hole that has been punched into a piece of material.

Ground Refers to electricity's habit of seeking the shortest route to earth. Neutral wires carry it there in all circuits. An additional grounding wire or the sheathing of the metal-clad cable or conduit protects against shock if the neutral leg is interrupted.

Ground iron The plumbing drain and waste lines that are installed beneath the basement floor. Cast iron was once used, but black plastic pipe (ABS) is now widely used.

Ground wire An electrical conductor that leads to an electric connection at the earth.

Groundwater Water from an aquifer or subsurface water source. The level of this water in the soil is called the "water table."

Grout A wet mixture of cement, sand, and water that flows into masonry or ceramic crevices to seal the cracks between the different pieces. Mortar made of such consistency (by adding water) will flow into the joints and cavities of the masonry work and fill them solid.

Guarded (electrical) Covered, shielded, fenced, enclosed, or otherwise protected by means of suitable covers, casings, barriers, rails, screens, mats, or platforms to remove the likelihood of approach or contact by persons or objects to a point of danger.

Gun nails Most modern framers use nail guns and rarely use a hammer. Gun nails come in strips or coils so they can be easily loaded in the nail gun. See **box nails**, **common nails**, and **sinkers**.

Gusset A flat wood, plywood, or similar type member used to provide a connection at the intersection of wood members. Most commonly used at joints of wood trusses. They

are fastened by nails, screws, bolts, or adhesives.

Gutter A shallow channel or conduit of metal or wood set below and along the (fascia) eaves of a house to catch and carry off rainwater from the roof.

Gutter brace A place where the gutter is attached to the structure's fascia.

Gyp board, drywall, wall board, and gypsum board A panel (normally 4 ft. × 8 ft., 10 ft., 12 ft., or 16 ft.) made with a core of gypsum (chalklike) rock that covers interior walls and ceilings.

Gypsum-based underlayment An underlayment made from fiber-reinforced gypsum that is easy to cut and install and is highly resistant to indentation. See **cement board underlayment**, **lauan plywood underlayment**, **particleboard underlayment**, **plywood underlayment**, **underlayment**, and **untempered hardboard underlayment**.

Gypsum plaster Gypsum formulated to be used with the addition of sand and water for base-coat plaster.

Habitable space A space in a building for living, sleeping, eating, and cooking. Most codes do not consider bathrooms, toilet rooms, closets, halls, storage, utility spaces, and so on as habitable spaces.

Half-bond A course of brick in which the vertical joint between bricks is halfway across the length of the brick in the course below it; also referred to as running bond and stretcher bond.

Half-laced valley Pattern formed in the valley of a roof by overlapping the valley with shingles from one side of the valley and cutting shingles from the other side so they end at the center of the valley.

Halogen burner Electrical burner that instantly becomes hot when the burner is turned on. See **ceran top**.

Halons Substances used in fire-suppression systems and fire extinguishers. These substances deplete the stratospheric ozone layer and are outlawed in some areas.

Hand/feet rule A rule for safe use of a ladder. The hand/feet rule states that when climbing a ladder you should always have either one foot and two hands or two feet and one hand on the ladder at all times.

Hand-split and resawn shake Wood shake with a rough, split face and a sawn back.

Hand texture Any texture that is applied to drywall by hand, without the use of a machine. This includes brush textures and hock textures.

Hanger wires Wire employed to suspend the acoustical ceiling from the existing structure. The standard material is 12-gauge, galvanized, soft-annealed steel wire, but heavier-gauge wire is available for higher load-carrying installations or situations where hanger wire spacing exceeds 4 ft. on center. Stainless steel wire and nickel-copper alloy wire are often used in severe-environment designs. Seismic designs or exterior installations subject to wind uplift may require supplemental bracing or substantial hanger devices such as metal straps, rods, or structural angles.

Hardscape Consists of the inanimate elements of the building landscaping, including pavement, roadways, stone walls, concrete paths, sidewalks, and concrete, brick, or tile patios.

Hardware All of the "metal" fittings that go into the home when it is near completion. For example, door knobs, towel bars, handrail brackets, closet rods, house numbers, door closers, and so on. Usually the interior trim carpenter installs the "hardware."

Haunch An extension and knee-like protrusion of the foundation wall that a concrete porch or patio will rest upon for support.

Hazard insurance Protection against damage caused by fire, windstorms, or other common hazards. Many lenders require borrowers to carry it in an amount at least equal to the mortgage.

H Clip Small metal clips formed like an "H" that fit at the joints of two plywood (or wafer board) sheets to stiffen the joint. Normally used on the roof sheeting.

Header A beam placed perpendicular to joists and to which joists are nailed in framing for a chimney, stairway, or other opening.

Header A wood lintel.

Header The horizontal structural member over an opening (e.g., over a door or window).

Header jamb Specific name for the jamb found on the top of the inside of window and door openings.

Hearth The fireproof area directly in front of a fireplace. A hearth is the inner or outer floor of a fireplace.

Heartwood Wood found at the center of the tree. Generally heartwood is higher-quality wood than sapwood, with less and tighter knots, and is more resistant to decay.

Heat blister A bubble that forms in a shingle when the asphaltic coating does not properly bond to the mat.

Heat exchanger A device that transfers heat from a source, such as a flame, to a conductor, such as air or water.

Heating degree-days The sum, on an annual basis, of the difference between 65°F and the mean temperature for each day as determined from "NOAA (National Oceanic and Atmospheric Administration) Annual Degree Days to Selected Bases Derived from the 1960–1990 Normals" or other weather data sources acceptable to code officials.

Heating load The amount of heating required to keep a building at a specified temperature during the winter, usually 65°F, regardless of outside temperature.

Heating zone Room or group of rooms that is heated or cooled as a unit, usually controlled through a single thermostat.

Heat island effect Refers to the absorption of heat by hardscapes, such as dark, nonreflective pavement and buildings, and its radiation to surrounding areas. Particularly in urban areas, other sources may include vehicle exhaust, air-conditioners, and street equipment; reduced airflow from tall buildings and narrow streets exacerbates the effect.

Heat line A distinct line left on walls by superheated smoke that was stopped at the ceiling. The bottom edge of this superheated smoke often leaves a line on the walls.

Heat meter An electrical municipal inspection of the electric meter breaker panel box.

Heat pump system A mechanical system that uses compression and decompression of gas to either take heat out of the structure or bring heat into the structure.

Heat rough Work performed by the heating contractor after the stairs and interior walls are built. This includes installing all ductwork and flue pipes. Sometimes, the furnace and fireplaces are installed at this stage of construction.

Heat trim Work done by the heating contractor to get the home ready for the municipal final heat inspection. This includes venting the hot water heater; installing all vent grills, registers, and air-conditioning services; turning on the furnace; installing thermostats; venting ranges and hoods; and all other heat-related work.

Heel cut A notch cut in the end of a rafter to permit it to fit flat on a wall and on the top, doubled, exterior wall plate.

Herringbone wood floor installation An installation where wood strips are installed in a zigzag pattern. See **diagonal wood floor installation** and **straight wood floor installation**.

High-density urethane foam pad A urethane foam pad that looks much like a thin wrestling pad. Unlike most other types of pad, water can be extracted from high-density urethane foam pads. See **rebound pad, synthetic felt pad**, and **waffle-type sponge rubber**.

High-efficiency particulate air (HEPA) filters Filters that remove virtually all (99.97%) 0.3-micron particles.

High-gloss paint As the name suggests high-gloss paint has a high gloss. High-gloss paint can be water based (also called latex or acrylic) or alkyd based, commonly referred to as oil-based paint. High-gloss alkyd paints are very hard, scrubbable, and resistant to abrasions.

Highlights A light spot, area, or streak on a painted surface.

Hip A roof with four sloping sides. The external angle formed by the meeting of two sloping sides of a roof.

Hip jack Type of jack rafter that runs from the hip rafter to the wall top plate.

Hip rafter A rafter that forms the hip line of the roof from the ridge to the outside corner of the exterior walls.

Hip roof A roof that rises by inclined planes from all four sides of a building.

Hoistway Any shaftway, hatchway, well hole, or other vertical opening or space in which an elevator or dumbwaiter is designed to operate.

Hold-down Used to connect the outside of the framing to the foundation. The hold-downs are placed in the foundation while the concrete is still wet.

Holidays Bubbles that result from the separation of plies in multiple-ply membrane roofs, usually due to improper installation.

Home run (electrical) The electrical cable that carries power from the main circuit breaker panel to the first electrical box, plug, or switch in the circuit.

Honeycombs The appearance concrete makes when rocks in the concrete are visible and where there are void areas in the foundation wall, especially around concrete foundation windows.

Hopper window A window unit that opens by moving the top of the window sash inward. The bottom of the window sash is attached with hinges.

Hose bib An exterior water faucet (sill cock).

Hot mop installation method Installation method where bonding materials are heated and mopped onto roofing materials to form a bond between layers, overlapping seams, or flashing. On modified bitumen roof systems, a method whereby styrene butadiene styrene–type (SBS-type) modified bitumen roofing is adhered to the base sheet.

Hot wire The wire that carries electrical energy to a receptacle or other device—in contrast to a neutral, which carries electricity away again. Normally it is the black wire. See **ground**.

Humidifier An appliance normally attached to the furnace, or a portable unit device designed to increase the humidity within a room or a house by means of the discharge of water vapor.

Hurricane clip Metal straps that are nailed to secure the roof rafters and trusses to the top horizontal wall plate; sometimes called a Teco clip.

Hurricane tie Manufactured metal bracket used to tie the roof truss to the top of the bearing wall.

HVAC An abbreviation for heat, ventilation, and air conditioning.

Hydration The chemical process that occurs when water and cement combine to form the adhesive paste that holds the aggregate together and makes concrete harden. The correct mixture of water, cement, and temperature is needed for proper hydration to occur.

Hydraulic shock The instantaneous pressure caused when a closed plumbing valve stops flowing water.

Hydrochlorofluorocarbons (HCFCs) Refrigerants that are used in building equipment and cause depletion of the stratospheric ozone layer but are less damaging than CFCs.

Hydrofluorocarbons (HFCs) Refrigerants that do not deplete the stratospheric ozone layer but may have high global warming potential and thus are not environmentally benign.

Hydrology The branch of geology that studies water on the earth and in the atmosphere and its distribution, uses, and conservation.

Hypotenuse The long side on a right triangle found opposite the 90-degree angle.

I-beam A steel beam with a cross section resembling the letter I. It is used for long spans as basement beams or over wide wall openings, such as a double garage door, when wall and roof loads bear down on the opening.

Ice damming A condition that occurs when snow melts on the heated portion of an improperly ventilated roof. The water drips down to the unheated portion where it freezes into ice. Eventually the buildup of ice will cause the roof to leak.

Identification The third of the six points of estimation of a loss. Identification is where the estimator decides what an item is. For example, identification occurs when an estimator decides that Formica will be used on countertops to replace the loss.

I-joist Manufactured structural building component resembling the letter I. Used as floor joists and rafters. I-joists include two key parts: flanges and webs. The flange of the I-joist may be made of laminated veneer lumber or dimensional lumber, usually formed into a 1½-in. width. The web or center of the I-joist is commonly made of plywood or oriented strand board (OSB). Large holes can be cut in the web to accommodate duct work and plumbing waste lines. I-joists are available in lengths up to 60 ft. long.

Imperviousness Resistance to penetration by a liquid, calculated as the percentage of an area covered by a paving system that does not allow moisture to soak into the ground.

Impervious surface Promotes runoff of precipitation instead of infiltration into the subsurface. The imperviousness or degree of runoff potential can be estimated for different surface materials.

Incandescent lamp A lamp employing an electrically charged metal filament that glows at white heat; a typical light bulb.

Index The interest rate or adjustment standard that determines the changes in monthly payments for an adjustable-rate loan.

Indirect cost Cost that can be identified with projects but not any specific project or unit of production.

Indoor air quality (IAQ) The nature of air that affects the health and well-being of building occupants.

Indoor-outdoor carpet A type of carpet that may be used in interior or exterior applications. Originally, indoor-outdoor carpet was made to imitate grass, but today it imitates many types of traditional interior carpet. See **berber carpet, nylon carpet, sculptured carpet, shag carpet,** and **wool carpet**.

Infiltration The passage of air from indoors to outdoors and vice versa; this term is usually associated with drafts from cracks, seams, or holes in buildings.

Infrared emittance Infrared emittance indicates the ability of a material to shed infrared radiation, whose wavelength is roughly 5 to 40 micrometers. Infrared emittance is measured on a scale of 0 to 1.

Injected insulation Insulation that is injected into place. There are two types of injected insulation: (1) foam that is injected into holes and cracks through tubes and (2) insulation that is injected through mesh into a framing cavity.

Inside corner The point at which two walls form an internal angle, as in the corner of a room.

Insulated flue A flue consisting of an inner pipe and an outer pipe with the space between the two filled with heat-resistant insulation.

Insulating glass Window or door in which two panes of glass are used with a sealed air space between; also known as double glass.

Insulation board, rigid A structural building board made of coarse wood or cane fiber in $\frac{1}{2}$- and $\frac{25}{32}$-in. thicknesses. It can be obtained in various size sheets and densities.

Insulation Any material high in resistance to heat transmission that, when placed in the walls, ceiling, or floors of a structure, will reduce the rate of heat flow.

Insulation stop Material placed in the roof system to prevent insulation from falling through the space between the top of the exterior wall and the bottom of the roof sheathing.

Integral drawer A type of drawer in which the face serves as the front piece of the drawer. The drawer sides attach directly to the drawer face. See **attached drawer**.

Integrated pest management (IPM) The coordinated use of knowledge about pests, environment, and pest prevention and control methods to minimize pest infestation and damage by the most economical means while minimizing hazards to the people, the property, and the environment.

Interest The cost paid to a lender for borrowed money.

Interior finish Material used to cover the interior framed areas of walls and ceilings.

Interior plaster Plaster that is used as a wall or ceiling finish inside the structure.

Interlocking shingle Shingles with interlocking edges; designed so that the wind cannot lift them. The most common type of interlocking shingle is the T-lock.

Interlocking siding Siding made from metal or vinyl with edges that interlock as the pieces are installed forming a weather-tight seam.

Interrupting rating The highest current at rated voltage that a device is intended to interrupt under standard test conditions.

Invasive plants Both indigenous and exotic species that are characteristically adaptable and aggressive, have a high reproductive capacity, and tend to overrun an area. Collectively, they are one of the great threats to biodiversity and ecosystem stability.

Irrigated land An area that has water delivered through artificial methods (i.e., other than rain). Conventional methods utilize pressure to deliver and distribute water through sprinkler heads. Other more efficient methods include drip irrigation, a highly efficient type of microirrigation that delivers water at a low pressure through buried mains and submains of perforated tubes or emitters.

Irrigation Lawn sprinkler or other similar system.

Isolated (as applied to location) Not readily accessible to persons unless special means for access are used.

Isolation membrane A protective layer of material that is installed between dissimilar materials. In a tile floor, an isolation membrane protects the tile from movement as the underlying system absorbs water that penetrates the tile surface. See **cementuous board**, **mortar bed**, and **thin-set tile**.

Jack post A type of structural support made of metal, which can be raised or lowered through a series of pins and a screw to meet the height required; basically used as a replacement for an old supporting member in a building. See **monopost**.

Jack rafter A rafter that spans the distance from the wall plate to a hip or from a valley to a ridge.

Jalousie window A window unit with numerous glass louvers that pivot simultaneously outward from the bottom.

Jamb The side and head lining of a doorway, window, or other opening. It includes studs as well as the frame and trim.

J Channel Metal edging used on drywall to give the edge a finished appearance when a wall is not "wrapped." Generally basement stairway walls have drywall only on the stair side. J Channel is used on the vertical edge of the last drywall sheet.

Joint The location between the touching surfaces of two members or components joined and held together by nails, glue, cement, mortar, or other means.

Joint cement or joint compound A powder that is usually mixed with water and used for joint treatment in gypsum-wallboard finish; often called "spackle" or drywall mud.

Joint tenancy A form of ownership in which the tenants own a property equally. If one dies, the other automatically inherits the entire property.

Joint trench When the electric company and telephone company dig one trench and "drop" both of their service lines in.

Joist Wooden 2 ft. × 8 ft., 10 ft., or 12 ft. beams that run parallel to one another and support a floor or ceiling and are supported in turn by larger beams, girders, or bearing walls.

Joist hanger A metal U-shaped item used to support the end of a floor joist and attached with hardened nails to another bearing joist or beam.

Jumpers Water pipe installed in a water meter pit (before the water meter is installed) or electric wire that is installed in the electric house panel meter socket before the meter is installed. This is sometimes illegal.

Junction box A box that protects splices in electrical wires and provides access. Switches, outlets, and boxes for light fixtures are junction boxes.

Keeper The metal latch plate in a door frame into which a doorknob plunger latches.

Keyless A plastic or porcelain light fixture that operates by a pull string. Generally found in the basement, crawl space, and attic areas.

Keyway A slot formed and poured on a footer or in a foundation wall when another wall will be installed at the slot location. This gives additional strength to the joint or meeting point.

Kiln drying A lumber-drying process where lumber is placed in a kiln or oven and heated until the excess moisture is removed. Kiln drying is about 10 times faster than surface drying.

Kilowatt (kw) One thousand watts. A kilowatt-hour is the base unit used in measuring electrical consumption.

King stud The vertical "two-bys" frame lumber (left and right) of a window or door opening and runs continuously from the bottom sole plate to the top plate.

Knob and tube wiring The wiring system used before 1945. Two strands of copper wire are run along framing by connecting them to porcelain knobs and through framing inside porcelain tubes.

Knockdown texture Any type of drywall texture that is flattened or smoothed. Texture may be knocked down when it is semidry with a drywall knife, or it can be sanded after it is dry.

Knot In lumber, the portion of a branch or limb of a tree that appears on the edge or face of the piece.

Kraft paper A heavy brown paper made of a sulfate pulp that is often used to face batt insulation. Kraft paper is not resistant to fire, so Kraft-faced batt insulation must be covered with a fire-resistant material.

Labeled Equipment or materials to which has been attached a label, symbol, or other identifying mark of an organization that is acceptable to the authority having jurisdiction and concerned with product evaluation, that maintains periodic inspection of production of labeled equipment or materials, and by whose labeling the manufacturer indicates compliance with appropriate standards or performance in a specified manner.

Laced valley Interlocking pattern formed in the valley of a roof by overlapping shingles in alternating rows, making a basket-weave pattern.

Lamella Shell construction in which the shell is formed by a lattice of interlacing members.

Laminated shingles Shingles that have extra layers or tabs, giving a shake-like appearance; may also be called "architectural shingles" or "three-dimensional shingles."

Laminated square edge (countertop) The edge on a countertop that is made by covering the square front corner with plastic laminate. See **plastic laminate countertop**.

Laminating Bonding together two or more layers of materials.

Lamp life The useful operating life of a lamp.

Lamps Lamps use electricity to produce light in any of several ways: by heating a wire for incandescence, by exciting a gas that produces ultraviolet light from a luminescent material, by generating an arc that emits visible light and some ultraviolet light, or by inducing excitation of mercury through radio frequencies. Light-emitting diodes, packaged as traditional bulbs, also fall under this definition.

Landfills Waste disposal sites for solid waste from human activities.

Landing A platform between flights of stairs or at the termination of a flight of stairs; often used when stairs change direction. Many building codes set minimum dimensions for landings.

Landscape area The project site area less the building footprint, hardscape areas, water bodies, and so on.

Lap To cover the surface of one shingle or roll with another.

Lap siding Horizontal siding that is installed by overlapping the top edge of each course with the bottom edge of the course directly above it.

Laser A tool used in building construction that projects a light beam out on a level plane for use as a reference in aligning objects so they are level (e.g., grade stakes and ceiling tiles).

Latch A beveled metal tongue operated by a spring-loaded knob or lever. The tongue's bevel lets you close the door and engage the locking mechanism, if any, without using a key; contrasts with **dead bolt**.

Lateral (electric, gas, telephone, sewer, and water) The underground trench and related services (i.e., electric, gas, telephone, sewer, and water lines) that will be buried within the trench.

Lath A building material of narrow wood, metal, gypsum, or insulating board that is fastened to the frame of a building to act as a base for plaster, shingles, or tiles.

Lattice An open framework of crisscrossed wood or metal strips that form regular patterned spaces.

Lauan plywood underlayment A plywood underlayment made from lauan wood. See **cement board underlayment**, **gypsum-based underlayment**, **particleboard underlayment**, **underlayment**, **plywood underlayment**, and **untempered hardboard underlayment**.

Laundry room pan A pan that is placed under a washer to catch water should the washer overflow. The pan may catch and hold the water or channel the water into a drain.

Layout board A board that has been marked to show the distance between each of the trusses. It is used while trusses are being installed to ensure they are positioned properly.

Lazy susan Type of specialty shelves that revolve.

Lead Electrical conductor.

Leader A vertical drainage pipe for conveying stormwater from roof or gutter drains to a building house storm drain, building house drain (combined), or other means of disposal. The leader includes the horizontal pipe to a single roof drain or gutter drain.

Ledger (for a structural floor) The wooden perimeter frame lumber member that bolts onto the face of a foundation wall and supports the wood structural floor.

Ledger strip A strip of lumber nailed along the bottom of the side of a girder on which joists rest.

Leech field A method used to treat and dispose of sewage in rural areas not accessible to a municipal sewer system. Sewage is discharged into a leech field.

LEED A system established by the US Green Building Council (USGBC) to define and measure "green" buildings.

Lessee The person in possession of a building under a lease from the building's owner.

Let-in brace Nominal 1-in.-thick boards applied into notched studs diagonally; also, L-shaped, metal straps that are installed by the framer at the rough stage to give support to an exterior wall or wall corner.

Level True horizontal; also a tool used to determine level.

Level cut A true horizontal cut. The actual angle required for a level cut is determined by the slope of the member to be installed. A level cut is at right angles to the plumb cut.

Level payment mortgage A mortgage with identical monthly payments over the life of the loan.

L-flashing L-shaped flashing that is made from one continuous piece of metal.

License (construction and construction related) A written document (usually) issued by a governmental agency authorizing a person to perform specific acts in or in connection with the construction or alteration of buildings or the installation, alteration, and use and operation of service equipment therein.

Lien An encumbrance that usually makes real or personal property the security for payment of a debt or discharge of an obligation.

Life-cycle assessment A technique to assess the environmental aspects and potential effects associated with a product, process, or service, by (1) compiling an inventory of relevant energy and material inputs and environmental releases, (2) evaluating the potential environmental effects associated with identified inputs and releases, and (3) interpreting the results to help make a more informed decision.

Light Space in a window sash for a single pane of glass; also, a pane of glass.

Lighting outlet An outlet intended for the direct connection of a lampholder, a luminaire (lighting fixture), or a pendant cord terminating in a lampholder.

Light pollution Waste light from building sites that produces glare, is directed upward to the sky, or is directed off the site. Waste light does not increase nighttime safety, utility, or security and needlessly consumes energy and natural resources. Beware of simple contract requirements that say "contractor to provide security lighting" or words to this effect, as light pollution may result in lawsuits from the local community.

Limit switch A safety control that automatically shuts off a furnace if it gets too hot.

Linear foot A 12-in. ruler is a linear ft. Some people use the term "linear ft.," to distinguish between a sq. ft. or a cu. ft. For example, if the room is 10 × 10 (linear ft.), it is 100 sq. ft.

Line loss (1) The reduction in the quantity of natural gas flowing through a pipeline that results from leaks, venting, and other physical and operational circumstances on a pipeline system, including leaks, theft, or fuel used by compressors to maintain pressure necessary for transportation. (2) A voltage drop caused by resistance in wire during transmission of electrical power over distance.

Linoleum Genuine linoleum, not to be confused with vinyl, was invented nearly 150 years ago and is still relevant today. Environmentally preferred linoleum is made from natural, raw materials. Linseed oil that comes from the flax plant is the primary ingredient. Other ingredients include wood or cork powder, resins, and ground limestone. Mineral pigments provide its colors. Sheet linoleum is available in many thicknesses.

Lintel A horizontal structural member that supports the load over an opening such as a door or window.

Listed Equipment, materials, or services included in a list published by an organization that is acceptable to the authority having jurisdiction; that is concerned with evaluation of products or services; that maintains periodic inspection of production of listed equipment or materials or periodic evaluation of services; and whose listing states that the equipment, material, or services either meets appropriate designated standards or has been tested and found suitable for a specified purpose. Note: Some owners may require equipment to be both listed and labeled.

Live load A temporary weight that will be placed on the structural part, including snow, people, and furniture; also, all occupants, materials, equipment, constructions, or other elements of weight supported in, on, or by a building that will or are likely to be moved or relocated during the expected life of the building.

Live parts (electrical) Energized conductive components.

Load-bearing wall Load bearing walls include all exterior walls and any interior wall that is aligned above a support beam or girder.

Loan The amount to be borrowed.

Loan-to-value (LTV) ratio The ratio of the loan amount to the property valuation, expressed as a percentage. For example, if a borrower is seeking a loan of $200,000 on a property worth $400,000, it has a 50% loan-to-value rate. The higher the loan to value, the greater the lender's perceived risk. Loans above normal lending LTV ratios may require additional security.

Lookout A short wood bracket or cantilever that supports an overhang portion of a roof.

Loop pile Carpet pile in which fibers are looped and both ends are attached to the carpet backing. See **cut pile** and **pile**.

Louver A vented opening into the home that has a series of horizontal slats and arranged to permit ventilation but to exclude rain, snow, light, insects, or other living creatures.

Low E A membrane that is placed between double panes of glass to filter the light coming through the window. It transfers more heat through the glass in the winter and blocks heat in the summer.

Lower unit (cabinet) A cabinet unit that is designed to sit on the floor. Also called a base unit. See **full-height cabinet, upper unit,** and **vanity cabinet**.

Low-profile water closet A water closet with a short tank that cannot usually be detached from the bowl. See **turbo toilet** and **water closet**.

Low-voltage wiring Wiring commonly used for television antennas, doorbells, thermostats, intercoms, and some specialty lighting systems.

Lumens A unit of measurement of the amount of brightness that comes from a light source. Lumens define "luminous flux," which is energy within the range of frequencies we perceive as light. For example, a wax candle generates 13 lumens; a 100-watt bulb generates 1200 lumens.

Luminaire A complete lighting unit consisting of a housing; lamp(s); light-controlling elements; brightness-controlling element; lampholder(s); auxiliary equipment, such as a ballast or transformer, if required; and a connection to a power supply.

L-V-L Beam Laminated veneer lumber or microlaminated beam, made from thin layers of wood, called veneers, that are glued together. The veneer segments may run either perpendicular or parallel to the load and they have no arch or camber.

Machine texture Any texture that is applied to drywall using a machine.

Main disconnect A set of large switches or breakers that allow electricity to the structure to be turned off without removing the meter. On newer structures the main disconnect is located on the meter base.

Makeup water Water fed into the system to replace what is lost through evaporation, drift, blowdown, and other causes.

Mansard roof A roof style with four sides similar to a **hip roof**, but each side is divided into an upper and lower section, the lower section having a steeper slope than the upper section. Often, the center of the Mansard roof consists of a flat roof.

Mantel The shelf above a fireplace opening. Also used in referring to the decorative trim around a fireplace opening.

Manufactured wood A wood product such as a truss, beam, gluelam, microlam, or joist that is manufactured out of smaller wood pieces and glued or mechanically fastened to form a larger piece. Often used to create a stronger member that may use less wood. See **oriented strand board**.

Manufacturer's specifications The written installation and/or maintenance instructions that are developed by the manufacturer of a product and that may have to be followed in order to maintain the product warrantee.

Masking The process of covering part of a surface that you do not want to paint during the present application. Masking involves placing tape on materials adjacent to the surface being painted to keep them clean. Typically masking is placed on trim that will be painted or stained a different color than the surface to which it is attached.

Masonry Stone, brick, concrete, hollow-tile, concrete block, or other similar building units or materials. Normally bonded together with mortar to form a wall.

Mass-produced cabinets Milled cabinets that are built in large quantities and in standard sizes. They are sold in high volume through retailers. See **built-in cabinets, custom cabinets**, and **milled cabinets**.

Mastic A pasty material used as a cement (as for setting tile) or a protective coating (as for thermal insulation or waterproofing).

Material safety data sheets (MSDS) MSDS contain product information on chemicals, chemical compounds, and chemical mixtures. MSDSs can also include instructions for safe handling, storage, and disposal of products.

MCM (microcircular mills) American standard gauge unit of measure for wire sizes that are larger than 4 aught. See **American standard gauge**.

MDF (medium density fiberboard) A type of pressed fiberboard often used in cabinet building.

Measuring tape A basic tool used to measure the length of materials. A measuring tape has marks on it that make it possible to read measurements with accuracy of up to 1/16 of 1 in. or more. Measuring tapes often have special marks at each 16-in. interval for easy location of 16-in.-on-center framing members.

Mechanical ventilation Ventilation provided by machine-powered equipment such as motor-driven fans and blowers but not by devices such as wind-driven turbine ventilators and mechanically operated windows.

Mechanics lien A lien on real property, created by statue in many years, in favor of persons supplying labor or materials for a building or structure, for the value of labor or materials supplied by them. In some jurisdictions, a mechanics lien also exists for the value of professional services. Clear title to the property cannot be obtained until the claim for the labor, materials, or professional services is settled. Timely filing is essential to support the encumbrance, and prescribed filing dates vary by jurisdiction.

Membrane A solid sheet of waterproof material that covers an entire roof area.

Messenger cable On an overhead electrical drop, the messenger cable is one of the three intertwined cables in the drop wire that contains the neutral lead and carries the weight of the other leads.

Metal-enclosed power switchgear A switchgear assembly completely enclosed on all sides and top with sheet metal (except for ventilating openings and inspection windows) containing primary power circuit switching, interrupting devices, or both with buses and connections. The assembly may include control and auxiliary devices. Access to the interior of the enclosure is provided by doors, removable covers, or both.

Metal lath Sheets of metal that are slit to form openings within the lath. Used as a plaster base for walls and ceilings and as reinforcement over other forms of plaster base.

Metal stirrup Used to connect a wood column or post to a concrete part. It holds the wood member securely while preventing the long grains of the wood from directly contacting the moist concrete. This keeps the moisture that collects on the concrete from being pulled into the wood by capillary action.

Meter base The base into which the power meter is attached.

Methylmercury Any of various toxic compounds of mercury containing the complex CH_3Hg^+. It often occurs in pollutants and bioaccumulates in living organisms, especially in higher levels of the food chain.

Microlam A manufactured structural wood beam. It is constructed of pressure and adhesive bonded wood strands. They have a higher strength rating than solid sawn lumber.

Milar (mylar) Plastic, transparent copies of a blueprint.

Milled cabinets Cabinets built by a manufacturer in a cabinet mill. See **built-in cabinets**, **custom cabinets**, and **mass-produced cabinets**.

Mill finish The surface finish found on aluminum when it is extruded at the mill.

Millwork Generally all building materials made of finished wood and manufactured in millwork plants including all doors, window and door frames, blinds, mantels, panelwork, stairway components (balusters, rail, etc.), moldings, and interior trim. Millwork does not include flooring, ceiling, or siding.

Mineral fibers There are many mineral fibers. For example, asbestos is a mineral fiber that is no longer used because of its adverse effects on health. A common type of mineral fiber insulation is called rock wool.

Minimum efficiency reporting value (MERV) A filter rating established by the American Society of Heating, Refrigerating and Air Conditioning Engineers (ASHRAE). The MERV efficiency categories range from 1 (very low efficiency) to 16 (very high). Alteration projects frequently require that the HVAC system of the building be protected from construction dust with a MERV 8 filter.

Miter joint The joint of two pieces at an angle that bisects the joining angle. For example, the miter joint at the side and head casing at a door opening is made at a 45-degree angle.

Mixed-mode ventilation A combination of natural ventilation and mechanical ventilation that allows the building to be ventilated mechanically or naturally and at times, both simultaneously.

Modified bitumen roof system A single membrane roof system made from either asphalt or coal tar pitch with added plasticizers. Installation methods for modified bitumen roof systems include both hot mop and torch down.

Molding A wood strip having an engraved, decorative surface.

Monolithic pour A nonstop concrete pour. An example of monolithic pour is when the footings, foundation, and floor slab are all formed and then poured at the same time.

Monopost An adjustable metal column used to support a beam or bearing point.

Mono truss A truss that has only one slope so that its outline is a triangle. Generally installed to rest on an exterior wall and on an inside bearing wall or to bear on the vertical member at the high end of the truss. Mono trusses may be used to form the outside portion of a Mansard roof.

Mortar A mixture of cement (or lime) with sand and water used in masonry work.

Mortar bed A type of isolation membrane that is made by spreading a layer of mortar, usually between ½ to 1½ in. thick, over the substrate. After the mortar bed cures, the tiles are attached. See **cement board, thin-set tiles,** and **isolation membrane**.

Mortgage A loan that is usually secured by the building and land for which the loan is given.

Mortgage broker A broker who represents numerous lenders and helps consumers find affordable mortgages.

Mortgage company A company that borrows money from a bank, lends it to consumers to buy homes, then sells the loans to investors.

Mortgage deed A legal document establishing a loan on property.

Mortgagee The lender who makes the mortgage loan.

Mortgage loan A contract in which the borrower's property is pledged as collateral. It is repaid in installments. The mortgagor (buyer) promises to repay principal and interest, keep the home insured, pay all taxes, and keep the property in good condition.

Mortgage origination fee A charge for work involved in preparing and servicing a mortgage application.

Mortise A slot cut into a board, plank, or timber, usually edgewise, to receive the tenon (or tongue) of another board, plank, or timber to form a joint.

Motor control center An assembly of one or more enclosed sections having a common power bus and principally containing motor control units.

Mudsill The bottom horizontal member of an exterior wall frame that rests on top of a foundation; also known as a sill plate or sole plate.

Mullion A vertical divider in the frame between windows, doors, or other openings.

Multiple-ply membrane A roof system with more than one layer. Multiple-ply membrane roof systems are also called built-up roofs and hot tar roofs. They are usually made from roll roofing materials that are bonded together with asphalt. A three-ply roof has a base sheet, ply sheet, and cap sheet. A five-ply roof has a base sheet, three-ply sheets, and a cap sheet. Hot tar (asphalt) is used to bond the plies and make the roof watertight.

Muntin A small member that divides the glass or openings of sash or doors.

Muriatic acid Commonly used as a brick cleaner after masonry work is completed.

Mushroom The unacceptable occurrence when the top of a caisson concrete pier spreads out and hardens to become wider than the foundation wall thickness.

Nail gun See **pneumatic nailer** and **stud gun**.

Nail inspection An inspection made by a municipal building inspector after the drywall material is hung with nails and screws. This inspection is conducted prior to taping.

Native vegetation and adapted vegetation Plants that are indigenous to a locality or plants that are adapted to the local climate and are not considered invasive species or noxious weeds. Native vegetation and adapted vegetation require limited irrigation following planting; do not require active maintenance, such as mowing; and provide habitat value.

Natural break A naturally occurring transition in a material. For example, on walls natural breaks occur at corners or where one material such as painted walls intersects another type of material, such as wallpaper.

Natural finish A transparent finish that does not seriously alter the original color or grain of the natural wood. Natural finishes are usually provided by sealers, oils, varnishes, water repellent preservatives, and other similar materials.

NEC (National Electrical Code) A set of rules governing safe wiring methods. Local codes that are backed by law may differ from the NEC in some ways.

Net present worth, net present value The difference between the present value of cash inflows and the present value of cash outflows. NPV is used in capital budgeting to analyze the profitability of an investment or project.

Neutral conductor One of three electrical conductors provided to a residential structure. The neutral conductor is connected with one of the current carrying leads to provide 120-volt power.

Neutral wire Usually color-coded white, the neutral wire carries electricity from an outlet back to the service panel. See **ground** and **hot wire**.

Newel post The large starting post to which the end of a stair guard railing or balustrade is fastened.

Noise reduction coefficient (NRC) The average sound absorption coefficient measured at four frequencies: 250, 500, 1000, and 2000 Hz expressed to the nearest integral multiple of 0.05; rates the ability of a ceiling, wall panel, or other construction to absorb sound. NRC is the fraction of sound energy, averaged over all angles of direction and from low to high sound frequencies, that is absorbed and not reflected.

Nonautomatic (electrical) Action requiring personal intervention for its control. As applied to an electric controller, nonautomatic control does not necessarily imply a manual controller but only that personal intervention is necessary.

Nonbearing wall A wall supporting no load other than its own weight.

Nonlinear load (electrical) A load where the wave shape of the steady-state current does not follow the wave shape of the applied voltage. Electronic equipment, electronic and electric-discharge lighting, adjustable-speed drive systems, and similar equipment may be nonlinear loads.

Nonoccupied spaces Generally rooms used by maintenance personnel and not open to occupants. Examples include janitorial closets, cleaning supply storage, and equipment rooms. Local building and zoning codes may define occupied, nonoccupied, and nonregularly occupied spaces, and the like. These local definitions need to be read and understood as the net size and economic viability of a project may rest on them.

Nonstructural part The part of a building that is not essential for supporting a load or for keeping the structure intact.

Nosing The projecting edge of a molding or drip or the front edge of a stair tread.

Notch A crosswise groove at the end of a board.

Note A formal document showing the existence of a debt and stating the terms of repayment.

Notes Instructions placed on plans by the architect or engineer. They are an integral part of the contract documents and must be followed.

Nozzle The part of a heating system that sprays the fuel of fuel-air mixture into the combustion chamber.

Number The fourth of the six points of estimation of a loss. Number is where the estimator calculates how many units of an item need to be replaced. For example, number includes calculating things such as the cu. yds. of concrete in a slab or the sq. ft. of drywall on a ceiling.

Number one common oak grade The grade of oak strip flooring that may have bright spots of sap, pinworm holes, machine defects, streaky or inconsistent color, grain variations, and a few knots. See **number two common oak grade** and **select and better oak grade**.

Number two common oak grade The grade of oak strip flooring that may have pronounced bright spots of sap, pinworm holes, machine defects, streaky or inconsistent color, grain variations, and knots. Some defects found in number two common oak grade are so severe that the installer will want to cut out some bad spots or even discard some severely flawed strips of wood. See **number one common oak grade** and **select and better oak grade**.

Nylon carpet A carpet made from man-made nylon fibers. See **berber carpet, indoor-outdoor carpet,**

sculptured carpet, shag carpet, and **wool carpet**.

Oakum Oakum is loose hemp or jute fiber that is impregnated with tar or pitch. Oakum is used to caulk large seams or for packing plumbing pipe joints.

OC (on center) The measurement of spacing for studs, rafters, and joists in a building from the center of one member to the center of the next.

On-demand water heaters Water heaters that heat water only when it is needed and then apply only the amount of heating required to satisfy the user's immediate needs. Many utility companies offer sizable rebates for installing on-demand water heaters; also called instantaneous hot water heaters.

One-call An agency that will mark the location of all utility lines. This service is not available in all areas and the name of the agency varies from state to state.

One-dimensional calculations Items that are simply counted or measured in just one direction. "Each" and "linear ft." are examples of one-dimensional units of measure.

One-quarter bond A course of brick where the vertical joint between bricks is one fourth of the way across the length of the brick in the course below it.

One-quarter turn A handrail piece that is used to turn a 90-degree corner. See **balustrade, goose neck, top rail**, and **volute**.

One-third bond Course of brick where the vertical joint between bricks is one third of the way across the length of the brick in the course below it.

On-site salvaged materials Materials recovered from and reused at the same building site. Do not assume all materials that are in good condition can be reused. Check first with local codes.

Open-grid pavement Generally pavement is deemed to be open-grid when it is less than 50% impervious and contains vegetation in the open wells.

Open hole inspection When an engineer (or municipal inspector) inspects the open excavation and examines the earth to determine the type of foundation (caisson, footer, wall on ground, etc.) that should be installed in the hole.

Open valley Method of flashing the valley of a roof in which the corrosion-resistant flashing material is left exposed while shingles from each side overlap the edges of the flashing.

Orange peel texture Finish applied to drywall with a machine that splatters mud onto the walls leaving a bumpy texture that is similar to the pattern on an orange peel.

Order of operations The mathematic rules that specify the order in which mathematic operations must be accomplished to produce the correct result. The order of operations is as follows:

1. Operations inside parenthesis
2. Squares and square roots
3. Multiplication and division
4. Addition and subtraction

Organic shingles Composition shingles made with an organic mat.

Organization The second of the six points of estimation of a loss. Organization is the method used to document the loss. Organization often begins with an accurate and detailed diagram. Next, the loss is typically organized into interior and exterior areas. The interior is organized by levels, such as the basement, main floor, and attic. Each level is then broken down into rooms. The estimate should usually start in the room and level where the damage point of origin is found. Other rooms are estimated in a clockwise (or counterclockwise) manner.

Oriented strand board (OSB) A manufactured 4 ft. × 8 ft. wood panel made out of 1- to 2-in. wood chips

and glue; often used as a substitute for plywood.

OSHA (Occupational Safety and Health Administration) A federally funded agency in the Department of Labor that develops job safety and health standards. The states have parallel organizations (e.g., CAL/OSHA).

Outdoor air The ambient air that enters a building either through a ventilation system or by infiltration (ASHRAE 62.1-2004).

Outlet (electrical) A point on the wiring system at which a current is taken to supply utilization equipment.

Outrigger An extension of a rafter beyond the wall line. Usually a smaller member nailed to a larger rafter to form a cornice or roof overhang.

Outside corner The point at which two walls form an external angle—one you usually can walk around.

Overcurrent Any current in excess of the rated current of equipment or the ampacity of a conductor. It may result from an overload, short circuit, or ground fault.

Overhang An outward-projecting, eave-soffit area of a roof; the part of the roof that hangs out or over the outside wall. See also **cornice**.

Overhead (elevator) The upper portion of the elevator hoistway. Codes establish the required overhead. Overhead may rise above the roof line thereby affecting a building's aesthetics.

Overhead costs Costs that cannot be identified with or charged to projects or units of production except on some more or less arbitrary allocation basis. These are costs that are incurred in the home office.

Overhead door A door commonly found on garages, mounted in a track or frame enabling it to move above the opening when in the open position.

Overlap joint Siding joint made by placing the edge of a piece of siding over a previously installed piece of siding.

Overlapping seam A waterproof seam created by placing the edge of one long side of a metal roofing panel over the long edge of the panel adjacent to it. A gasket may be placed along the seam.

Overlaying shingles The process of laying a new layer of shingles on top of an old layer of shingles. Overlaying shingles can make a roof more susceptible to hail damage. Weight is also an issue of concern. Most building codes do not allow more than two layers of shingles.

Overload (electrical) Operation of equipment in excess of normal, full-load rating or of a conductor in excess of rated ampacity that, when it persists for a sufficient length of time, would cause damage or dangerous overheating. A fault, such as a short circuit or ground fault, is not an overload.

Overspray Fine particles of paint that are carried in the air from the paint sprayer. These particles may then land on other surfaces that are not intended to be painted.

Owner (1) An entity that possesses the exclusive right to hold, use, benefit from, enjoy, convey, transfer, and otherwise dispose of an asset or property; (2) a person or entity who awards a contract for a project and undertakes to pay the contractor, also called contract owner; and (3) an employee or executive who has the principle responsibility for a process, program, or project.

Padding A material installed under carpet to add foot comfort, isolate sound, and prolong carpet life.

Pad out, pack out To shim or add strips of wood to a wall or ceiling in order that the finished ceiling or wall will appear correct.

Paint A combination of pigments with suitable thinners or oils to provide decorative and protective coatings. Paints are oil based or latex water based.

Paint grade material Trim material that has flaws or joints that will be hidden if the material is painted.

Paint sprayer A high pressure paint pump that sprays paint through a sprayer tip or nozzle onto the surface. There are air-powered and airless paint sprayers.

Pallets Wooden platforms used for storing and shipping material. Forklifts and hand trucks are used to move these wooden platforms around.

Panel A thin, flat piece of wood, plywood, or similar material framed by stiles and rails as in a door (or cabinet door) or fitted into grooves of thicker material with molded edges for decorative wall treatment.

Panelboard A single panel or group of panel units designed for assembly in the form of a single panel, including buses and automatic overcurrent devices, and equipped with or without switches for the control of light, heat, or power circuits; designed to be placed in a cabinet or cutout box placed in or against a wall, partition, or other support and to be accessible only from the front.

Pan tile A U-shaped roofing tile that forms the troughs in a barrel tile roof.

Paper (building) A general term for papers, felts, and similar sheet materials used in buildings without reference to their properties or uses. Generally comes in long rolls.

Paper burns Tiny fibers that are raised in the drywall surface when sanded excessively.

Parallelogram A four-sided shape in which the diagonals drawn from opposite corners are not equal and opposite sides are parallel to each other.

Parapet wall Often found around flat roof systems, a parapet wall is a low wall that rises above the roof deck. Parapet walls are to prevent people from falling and to prevent roof fires from entering upper floor windows and vice versa.

Parquet flooring A wood floor that consists of small pieces of wood that are arranged into a specific design. Parquet flooring is generally made from premanufactured interlocking blocks. See **plank flooring**, **plug-and-plank flooring**, and **strip flooring**.

Partially shielded An exterior light fixture that is shielded so that the lower edge of the shield is at or below the centerline of the lamp to minimize light emitted above the horizontal plane.

Particle board A plywood substitute made of course sawdust that is mixed with resin and pressed into sheets. Used for closet shelving, floor underlayment, stair treads, and so on.

Parting stop or strip A small wood piece used in the side and head jambs of double hung windows to separate the upper sash from the lower sash.

Partition wall A wall that subdivides spaces within any story of a building or room.

Passage hardware Interior door hardware that does not lock. See **privacy hardware**.

Pattern layer (vinyl flooring) One of three layers of material typically found in vinyl floor covering. The pattern layer is the inner foam layer that is sandwiched between the backing and the wear layer.

Pattern rafter A rafter used as a guide for making all other common rafters.

Paver, paving Materials—commonly masonry—laid down to make a firm, even surface. Any tile or masonry unit that can be used as a surface upon which one may drive or walk.

Payment schedule A preagreed upon schedule of payments to a contractor usually based upon the amount of work completed. Such a schedule may include a deposit prior to the start of work. There may also be a temporary "retainer" (5–10% of the total cost of the job) at the end of the contract for correcting any small items that have not been completed or repaired.

PB (polybutylene) A type of plastic plumbing pipe made from polybutylene.

PB manifold system A system that distributes hot and cold water to individual plumbing fixtures from a single panel of valves.

Pea gravel Aggregate or stones roughly the size of peas.

Peak Highest part of the roof where the roof planes meet; also called the ridge.

Pedestal A metal box installed at various locations along utility easements that contain electrical, telephone, or cable television switches and connections.

Penalty clause A provision in a contract that provides for a reduction in the amount otherwise payable under a contract to a contractor as a penalty for failure to meet deadlines or for failure of the project to meet contract specifications.

Penny As applied to nails, it originally indicated the price per hundred. The term now serves as a measure of nail length and is abbreviated by the letter "d." Normally, 16d (16 "penny") nails are used for framing. See **common nails**.

Percent of completion method
An accounting method that recognizes revenues and gross profit each period based upon the progress of construction (i.e., the percentage of completion).

Percolation The general movement of liquid through soil.

Percolation test or perc. test
A test to determine how well water passes through soil. Commonly used to determine the feasibility of installing a leech field–type sewer system and for sizing such a leech field.

Performance bond An amount of money (usually 10% of the total price of a job) that a contractor must put on deposit with a governmental agency as an insurance policy that guarantees the contractors' proper and timely completion of a project or job.

Perimeter The measurement around the outside of an object. The perimeter of a rectangle or triangle is the sum of the lengths of its sides. On a circle the perimeter is also referred to as the circumference and is calculated by multiplying pi (3.14) and the diameter.

Perimeter backing A type of backing used on some types of specialty vinyl floor covering that is installed with adhesive placed only around the perimeter of the room. See **backing layer**.

Perimeter drain Usually 3- or 4-in. perforated plastic pipe that goes around the perimeter (either inside or outside) of a foundation wall (before backfill) and collects and diverts ground water away from the foundation. Generally it is "daylighted" into a sump pit inside the home, and a sump pump is sometimes inserted into the pit to discharge any accumulation of water.

Permeability A measure of the ease with which water penetrates a material.

Permit A governmental authorization to perform a building process as in the following:

- *Zoning/use permit.* Authorization to use a property for a specific use (e.g., a garage, a single family residence, etc.).

- *Demolition permit.* Authorization to tear down and remove an existing structure.

- *Grading permit.* Authorization to change the contour of the land.

- *Septic permit.* A health department authorization to build or modify a septic system.

- Building permit. Authorization to build or modify a structure.

- *Electrical permit*. A separate permit required for most electrical work.
- *Plumbing permit*. A separate permit required for new plumbing and larger modifications of existing plumbing systems.

And there are others that localities may require. It is not unusual for each permit to be independently issued. Therefore, as a contractor, you may have to spend considerable time going from one office to another to obtain all the permits necessary to begin a project.

Perspective The first of the six points of estimation of a loss. Perspective is where the estimator gains an understanding of the loss and decides where and how to proceed. Perspective includes such things as determining the type of structure, learning what caused the damage, noting subrogation issues, locating the damage point of origin, taking photos, and developing a theory of the total effect of the damaging event. Also, the estimator needs to determine if code changes have occurred since the original construction and whether or not the original work is grandfathered.

Perviousness The permeability of an area. A concrete driveway would be considered impervious while a lawn would be deemed pervious. See **percolation test or perc. test**.

Physical disability, physical handicap Do not assume that this definition is complete and accurate in all places. If disabilities are at issue, check your local laws before proceeding. A person is considered physically disabled if one or more of the following conditions exist: (a) impairment requiring use of a wheelchair; (b) impairment causing difficulty or insecurity in walking or climbing stairs or requiring the use of braces, crutches, or other artificial supports; (c) impairment caused by amputation, arthritis, spastic condition, or pulmonary, cardiac, or other ills rendering the individual semiambulatory; (d) total or partial impairment of hearing or sight causing insecurity or likelihood of exposure to danger in public places; or (e) impairment due to conditions of aging and incoordination.

Pi (Greek alphabet π) A mathematical constant used in many calculations involving circles. Pi is equal to 3.141596 . . . but is generally rounded to 3.14 for use in this course.

Picogram One trillionth of a gram.

Picograms per lumen-hour A measure of the amount of mercury in a lamp per unit of light delivered over its useful life. If you are responsible for purchasing mercury-containing lamps, seek to buy lamps that contain no more than 70 picograms of mercury per lumen-hour.

Pier A column, usually rectangular in horizontal cross section, used to support other structural members; also see **caisson**.

Pier-and-grade beam structure A structure utilizing piers and grade beams. It is built upon footings that have been constructed at a lower level to rest on bedrock or stable soil. This type of construction is commonly used on steep hillsides.

Pigment A powdered solid used in paint or enamel to give it a color.

Pigtails (electrical) The electric cord that the electrician provides and installs on an appliance such as a garbage disposal, dishwasher, or range hood.

Pile A long, slender column usually of timber, steel, or reinforced concrete driven into the ground to carry a vertical load.

Pile (carpeting) Carpet fibers that have been attached to a carpet backing. See **cut pile** and **loop pile**.

Pilot hole A small-diameter, predrilled hole that guides a nail or screw.

Pilot light A small, continuous flame (in a hot water heater, boiler, or furnace) that ignites gas or oil burners when needed. Pilot lights have been replaced by electric starters on new appliances.

Pipe dope Compound placed on threads that helps to seal threaded pipe joints.

Pipe flashing A flashing that is placed around any pipe that penetrates the roof. A gasket-like sleeve fits around the pipe, and its base slides under the upper shingle and over the lower shingle.

Pitch The inclined slope of a roof or the ratio of the total rise to the total width of a house (i.e., a 6-ft. rise and 24-ft. width is a one-fourth pitch roof). Roof slope is expressed in the in. of rise per ft. of horizontal run.

PITI Principal, interest, taxes, and insurance (the four major components of monthly housing payments).

Plain sawing Plain sawing or slash sawing is the most common method of sawing logs into lumber. This is really a matter of slicing the log into the required sizes.

Plan A set of pictures of whatever project is being worked on. A plan provides instruction for the construction of the project depicted.

Plank flooring Wood floor containing wood strips that are over 3¼ in. wide. See **parquet flooring**, **plug-and-plank flooring**, and **strip flooring**.

Plan view Drawing of a project, usually a structure, with the view from overhead, looking down.

Plaster Made from concrete, water, and aggregate, as are concrete, grout, and mortar. It is mixed using sand as its aggregate and using no course aggregate, such as gravel. It produces a hard, concrete-like surface. Plaster is often used as a finish coat on the exterior of a block wall.

Plasticiser Additive that increases the flexibility of a material.

Plasticity Refers to the softness of concrete, mortar, or soil or to how easy it is to mold and shape. Concrete that has a high-plasticity flows easily and is easy to work with. When concrete loses its plasticity, it becomes hard and can no longer be worked.

Plastic laminate countertop Type of countertop in which plastic laminate veneer is glued over supporting material that is usually made from plywood or medium density fiberboard (MDF). In practice, many people refer to plastic laminate by the brand name Formica. See **cultured countertop, cultured marble countertop, solid plastic countertop, solid surface countertop, stone countertop, tile countertop**, and **wood block countertop**.

Plate Normally a 2 in. × 4 in. or 2 in. × 6 in. that lays horizontally within a framed structure, such as the following:

- *Sill plate.* A horizontal member anchored to a concrete or masonry wall
- *Sole plate.* Bottom horizontal member of a frame wall
- *Top plate.* Top horizontal member of a frame wall that supports ceiling joists, rafters, or other members

Plate glass mirror A mirror made from high-quality glass that can be up to 1¼ in. thick. A plate glass mirror can be glued directly to the wall, held in a frame that is hung on the wall, or held in place with plastic clips.

Plenum The main hot air supply duct leading from a furnace. In a suspension ceiling construction, it is the space between the suspended ceiling and the main structure above.

Plot plan An overhead view plan that shows the location of the home on the lot. Includes all easements, property lines, setbacks, and legal descriptions of the home. Most building departments require a plot plan whenever you apply for a building permit.

Plough, plow To cut a lengthwise groove in a board or plank. An exterior handrail normally has a ploughed groove for hand gripping purposes.

Plug-and-plank flooring Wood flooring that is installed by installing fasteners through the tops of boards. Holes for fasteners are predrilled along with a countersink hole. After the fasteners are tightened into place, the holes are filled with plugs that are usually sanded level with the finish floor. See **parquet flooring**, **plank flooring**, and **strip flooring**.

Plumb Exactly vertical and perpendicular.

Plumb bob A lead weight attached to a string. It is the tool used in determining plumb.

Plumb cut True vertical cut when a member is installed. The actual angle required for a plumb cut is determined by the slope of the member. A plumb cut is at a right angle to the level cut.

Plumbing boots Metal saddles used to strengthen a bearing wall or vertical stud(s) where a plumbing drain line has been cut through and installed.

Plumbing fixtures and fittings Receptacles, devices, or appliances that are either permanently or temporarily connected to the building's water distribution system, receive liquid or liquid-borne wastes, and discharge wastewater, liquid-borne waste materials, or sewage either directly or indirectly to the drainage system of the premises. This includes water closets, urinals, lavatories, sinks, showers, and drinking fountains.

Plumbing ground The plumbing drain and waste lines that are installed beneath a basement floor.

Plumbing jacks Sleeves that fit around drain and waste vent pipes at, and are nailed to, the roof sheeting.

Plumbing rough Work performed by the plumbing contractor after the rough heat is installed. This work includes installing all drain and waste lines, copper water lines, bath tubs, shower pans, and gas piping to furnaces and fireplaces.

Plumbing stack A plumbing vent pipe that penetrates the roof.

Plumbing trim Work performed by the plumbing contractor to get the home ready for a final plumbing inspection; includes installing all toilets (water closets), hot water heaters, and sinks; and connecting all gas pipes to appliances, the disposal, dishwasher, and all plumbing items.

Plumbing waste line Pipe used to collect and drain sewage waste.

Ply A term to denote the number of layers of roofing felt, veneer in plywood, or layers in built-up materials, in any finished piece of such material.

Plywood A panel (normally 4 ft. × 8 ft.) of wood made of three or more layers of veneer, compressed and joined with glue, and usually laid with the grain of adjoining plies at right angles to give the sheet strength.

Plywood underlayment An underlayment made from plywood that has been specially manufactured for use as a vinyl floor underlayment. See **cement board underlayment**, **gypsum-based underlayment**, **lauan plywood underlayment**, **particleboard underlayment**, **underlayment**, and **untempered hardboard underlayment**.

Pneumatic nailer A tool that uses compressed air to drive a nail, usually into wood; it can also drive special nails into hard surfaces like concrete.

Pocket door A sliding door that rolls on a track and opens into a cavity in the wall.

Point load A point where a bearing or structural weight is concentrated and transferred to the foundation.

Point of origin The point from which the damage began. For example, in a fire, the point of origin is where the fire started.

Pollutants Emissions of carbon dioxide (CO_2), sulfur dioxide (SO_2), nitrogen oxides (NO_2), mercury (Hg), small particles (particulate matter, PM-2.5), and large particles (PM-10).

Polyvinyl acetate (PVA) Water-based primer commonly used on drywall. When used on wood, PVA can cause raised grains.

Popcorn texture Finish applied to drywall ceilings that contains large clumps of texturing material similar to popcorn or cottage cheese; also called acoustic texture or cottage cheese texture.

Portland cement Cement made by heating clay and crushed limestone into a brick and then grinding to a pulverized powder state.

Post A vertical framing member usually designed to carry a beam. Often a 4 in. × 4 in.; a 6 in. × 6 in.; or a metal pipe with a flat plate on top and bottom.

Post-and-beam A basic building method that uses just a few hefty posts and beams to support an entire structure; contrasts with stud framing.

Postconsumer content The percentage of material in a product that is recycled from consumer waste. LEED projects may specify postconsumer content.

Postconsumer fiber Paper, paperboard, and fibrous wastes that are collected from municipal waste streams.

Postconsumer material Recycled from consumer waste.

Postformed countertop Type of plastic laminate countertop that includes an integral rolled backsplash and a rolled front edge. Postformed countertops are fabricated in a shop. See **flat-laid countertop** and **plastic laminate countertop**.

Poured-in insulation Loose insulation that comes in bags and is poured into place. Most commonly used to fill cavities in masonry walls.

Power outlet An enclosed assembly that may include receptacles, circuit breakers, fuseholders, fused switches, buses, and watt-hour meter mounting means; intended to supply and control power to mobile homes, recreational vehicles, park trailers, or boats or to serve as a means for distributing power required to operate mobile or temporarily installed equipment.

Power roller Paint roller with a pump unit attached to bring the paint onto the roller so it can be spread evenly on the surface.

Power vent A vent that includes a fan to speed up air flow; often installed on roofs.

Preconsumer content Formerly known as postindustrial content, it is the percentage of material in a product that is recycled from manufacturing waste (e.g., sawdust). Excluded are materials such as rework, regrind, or scrap generated in a process and

capable of being reclaimed within the same process that generated it.

Prehung door A door unit that comes with the jamb assembled, door stop in place, and the slab connected to the jamb with hinges.

Premises wiring (system) The interior and exterior wiring, including power, lighting, control, and signal circuit wiring together with all their associated hardware, fittings, and wiring devices both permanently and temporarily installed, that extends from the service point or source of power, such as a battery, a solar photovoltaic system, or a generator, transformer, or converter windings, to the outlet(s). Such wiring does not include wiring internal to appliances, luminaires (fixtures), motors, controllers, motor control centers, and similar equipment.

Premium The amount payable on a loan.

Prepasted wallpaper Wallpaper with adhesive that has been applied to the paper then dehydrated. Prepasted wallpaper can be attached by dipping it into water and then placing it on the wall. However, in practice, many installers will apply a thin layer of paste.

Preservative Any pesticide that will prevent the action of wood-destroying fungi, insect borers, and similar destructive agents when the wood has been properly coated or impregnated with it; normally an arsenic derivative. Chromated copper arsenate (CCA) is an example.

Pressure relief valve (PRV) A device mounted on a hot water heater or boiler that is designed to release any high steam pressure in the tank to prevent tank explosions.

Pressure tank Used with a water pump to provide pressure in plumbing lines when the pump is not running. The pressure tank contains water and air. As water is pumped into the pressure tank, the air compresses. Then when a faucet is opened, the air pressure in the tank pushes the water through the lines. When pressure in the tank drops below a preset limit, the pump turns on and runs until the required pressure is reached.

Pressure-treated lumber Lumber that has been saturated with a preservative.

Pressurized irrigation water Water that is not fit for human consumption but is provided to a structure for use as irrigation water for plants and lawns. See also **graywater**.

Primer The first base coat of paint when a paint job consists of two or more coats. A first coating formulated to seal raw surfaces and hold succeeding finish coats.

Principal The original amount of the loan, the capital.

Privacy hardware Interior door hardware that locks. See **passage hardware**.

Process water Water used for industrial processes and building systems, such as cooling towers, boilers, and chillers.

Property survey A survey to determine the boundaries of your property. Some banks require a property survey prior to granting a mortgage.

Protected pattern A pattern left on a surface that was exposed to heat and smoke. Items on the surface, such as appliances and dishes, protect the areas they rest on from heat and smoke. When the surface is cleaned a pattern remains because the exposed areas discolor more than do the areas that were covered.

PSI Common abbreviation for pounds per sq. in.; unit of measure used to specify the strength of concrete.

P-trap A plumbing part, shaped like the letter P. A P-trap holds water that traps sewer gases in the line and prevents them from entering the structure.

Pull A handle used on drawers and cabinet doors.

Pulse system Furnace that produces heat through multiple explosions of gas.

Pump mix Special concrete that will be used in a concrete pump. Generally the mix has smaller rock aggregate than regular mix.

Punch list A list of discrepancies that need to be corrected by the contractor.

Punch out To inspect and make a punch list.

Purlin Horizontal member that spans across adjacent rafters or beams, commonly installed to provide a fastening surface for the roofing material.

Putty A type of dough used in sealing glass in the sash, filling small holes and crevices in wood, and for similar purposes.

PVC (polyvinyl chloride) or CPVC A type of plastic used to make elastomeric roof membranes as well as plumbing pipes, fittings, conduit, and fences.

Pythagorean theorem A mathematic relationship between sides of right triangles specifying that the length of the hypotenuse (longest side) squared equals the sum of the lengths of the other two sides squared. It is often expressed as $a^2 + b^2 = c^2$. The theorem can be used to solve for any side of a right triangle where the lengths of the other two sides are known. To solve for side a, use the formula $a^2 = c^2 - b^2$. To solve for side b, use the formula $b^2 = c^2 - a^2$.

Quad outlet An outlet with four ports.

Qualified person (electrical) One who has skills and knowledge related to the construction and operation of the electrical equipment and installations and has received safety training on the hazards involved.

Quarry tile A man-made or machine-made clay tile used to finish a floor or wall; generally 6 in. × 6 in. × ¼ in. thick.

Quarter round A small trim molding that has the cross section of a quarter circle.

Quartersawing A method of cutting boards from a log by sawing from the bark side of the log toward the center axis of the log. The method is called quartersawing because logs are usually split into quarters before they are sawn. Quartersawn boards have a consistent grain that runs at a 45- to 90-degree angle to the face of the board. See **plain sawing**.

Quoin Large stones, usually rectangular, used as decorative corners; generally installed to protrude or recess from the structure.

Rabbet A rectangular, longitudinal groove cut in the corner edge of a board or plank.

Rabbet joint A siding joint made by thinning the edge of two boards to about half their width and then overlapping the two thinned edges.

Raceway An enclosed channel of metal or nonmetallic materials designed expressly for holding wires, cables, or busbars, with additional functions as permitted in the NEC. Raceways include, but are not limited to, rigid metal conduit, liquidtight flexible conduit, electrical nonmetallic tubing, electrical metallic tubing, underfloor raceways, cellular concrete floor raceways, cellular metal floor raceways, surface raceways, wireways, and busways.

Radiant heating A method of heating, usually consisting of a forced hot water system with pipes placed in the floor, wall, or ceiling; also, electrically heated panels.

Radiation Energy transmitted from a heat source to the air around it. Radiators actually depend more on convection than radiation.

Radiator Heating device that transfers heat from water or steam running inside of it to the air and objects around it.

Radius A line extending from the center of a circle to the outside edge. It is equal to one-half the diameter of the circle.

Radon A naturally occurring, heavier than air, radioactive gas common in many parts of the country. Radon gas exposure is associated with lung cancer. Mitigation measures may involve crawl space and basement venting and various forms of vapor barriers.

Radon system A ventilation system beneath the floor of a basement or structural wood floor and designed to fan exhaust radon gas to the outside of the home.

Rafter Lumber used to support the roof sheeting and roof loads. Generally 2 in. × 10 in. and 2 in. × 12 in. rafters are used. The rafters of a flat roof are sometimes called roof joists.

Rafter (hip) A rafter that forms the intersection of an external roof angle.

Rafter (valley) A rafter that forms the intersection of an internal roof angle. The valley rafter is normally made of double, 2-in.-thick members.

Rail Cross members of panel doors or of a sash; also, a wall or open balustrade placed at the edge of a staircase, walkway bridge, or elevated surface to prevent people from falling off.

Rail Any relatively lightweight horizontal element, especially those found in fences (split rail).

Rail (cabinet) A horizontal member on the cabinet face frame. See **stile**.

Railroad tie Black tar and preservative impregnated, 6 in. × 8 in. and 6 ft. to 8 ft. long wooden timber that was used to hold railroad tracks in place; often used as a member of a retaining wall.

Rainproof Constructed, protected, or treated so as to prevent rain from interfering with the successful operation of an apparatus under specified test conditions.

Raintight Constructed or protected so that exposure to a beating rain will not result in the entrance of water under specified test conditions.

Raised panel A solid wood panel used in a frame and panel cabinet door. The edges of the panel are shaped to a thin edge so the panel will fit into the slots in the surrounding frame. See **flat panel**, **frame and panel cabinet door**, and **slab cabinet door**.

Raised panel door A door made from panels, usually framed and held in place by stiles and rails. Simulated raised panel doors are also made from pressed-wood fibers.

Rake Slope or slanted. Also, the edge The edge at the end of a sloping roof plane. For example, the roof edge at the top of the gable end wall is a rake.

Raked A mortar joint between courses of brick in which the mason uses a tool called a rake to remove the excess mortar to create a uniform depth.

Rake fascia The vertical face of the sloping end of a roof eave.

Rake siding The practice of installing lap siding diagonally.

Ranch A single-story, one-level home.

Rapidly renewable materials Agricultural products, both fiber and animal, that take 10 years or less to grow or raise and can be harvested in an ongoing and sustainable fashion.

Rasp Coarse file used to smooth and shape wood.

Ready-mixed concrete Concrete mixed at a plant or in trucks en route to a job and delivered ready for placement.

Ready-set tile A type of small mosaic tile that comes prefastened to a mat. Ready-set tiles are much faster to install than individual mosaic tile.

Rebar, reinforcing bar Ribbed steel bars installed in foundation concrete walls, footers, and poured in placed concrete structures designed to strengthen concrete. Comes in various thicknesses and strength grades.

Rebond pad A urethane foam carpet pad made by gluing small pieces of foam together. See **high-density urethane foam pad**, **synthetic felt pad**, and **waffle-type sponge rubber pad**.

Receptacle (electrical) An electrical outlet. A typical household will have many 120-volt receptacles for plugging in lamps and appliances and 240-volt receptacles for the range, clothes dryer, air-conditioners, and so on. A single receptacle is a single contact device with no other contact device on the same yoke. A multiple receptacle is two or more contact devices on the same yoke.

Receptacle outlet An outlet where one or more receptacles are installed.

Recess-mount medicine cabinet A medicine cabinet that is recessed into the wall, usually between two studs. See **surface-mount medicine cabinet**.

Recirculated air Air removed from a space and reused as supply air, delivered by mechanical or natural ventilation.

Reclaimed water Wastewater that has been treated and purified for reuse.

Recording fee A charge for recording the transfer of a property, paid to a city, county, or other appropriate branch of government.

Red label shingles Shingles made from high-grade wood with some slight sapwood and very little flat grain. Most residential structures use red or blue label shingles.

Redline (red-lined prints) Blueprints that reflect changes and that are marked with red pencil.

Reducer A fitting with different size openings at either end and used to go from a larger to a smaller pipe.

Reflective insulation Sheet material with one or both faces covered with aluminum foil.

Refrigerant gas A substance that remains a gas at low temperatures and pressure and can be used to transfer heat. Freon is an example and is used in air conditioning systems. Many refrigerant gasses are harmful to the environment and have been banned. Check carefully before purchasing AC equipment.

Regionally harvested (or extracted) and processed materials Materials that come from within a 500-mile radius of the project site.

Register A grill placed over a heating duct or cold air return.

Reglaze To replace a broken window.

Reinforcing steel Steel that is buried in concrete to provide added tensile strength. Usually called rebar, it contributes substantially to the strength of the concrete structural part.

Relative humidity The ratio of partial density of water vapor in the air to the saturation density of water vapor at the same temperature and the same total pressure.

Release film A thin strip of plastic attached to the underside of composition shingles to prevent the shingles from sticking together during shipment. When installed, the release film lines up with the sealant strip on the face of the shingle in the course below. When heated by the sun asphalt penetrates the release film and bonds to the sealant strip of the shingles in the underlying course.

Release powder Powder spread on the concrete surface before a concrete stamp is used. Prevents the concrete stamp from sticking to the concrete and adds a second color to the concrete.

Relief valve A device designed to open if it detects excess temperature or pressure.

Remote A remote electrical, gas, or water meter digital readout that is installed near the front of the home in order for utility companies to easily read the home owners usage of the service.

Remote-control circuit Any electrical circuit that controls any other circuit through a relay or an equivalent device.

Repairability The process of determining whether an item is repairable or should be replaced. If an item can be made to look and function the same as it did before the damaging event at a cost that is less than replacement cost, it should be repaired. If not, it should be replaced.

Retaining wall A structure that holds back a slope and prevents erosion. Many building codes require retaining walls over 4 ft. in height to be designed by a licensed engineer.

Retentions Amounts withheld from progress billings until final and satisfactory project completion.

Retrofit Any change to an existing facility, such as the addition or removal of equipment or a required adjustment, connection, or disconnection of equipment.

Return air Air removed from conditioned spaces and either recirculated in the building or exhausted to the outside.

Reverse board and batten Vertical siding in which narrow boards, called battens, are installed first with gaps between them. Wider boards are then installed over the gaps.

R-factor or value A measure of a material's resistance to the passage of heat. New home walls are usually insulated with 4 in. of batt insulation with an R-value of R-13 and a ceiling insulation of R-30.

Ribbon (girt) Normally a 1 in. × 4 in. board let into the studs horizontally to support the ceiling or second-floor joists.

Ridge The horizontal line at the junction of the top edges of two sloping roof surfaces.

Ridge board The board placed on the ridge of the roof onto which the upper ends of other rafters are fastened.

Ridge shingles Shingles that are used to cover the ridge board.

Ridge vent A vent placed along the ridge of the roof. It allows ventilation of the roof by raising the level of the ridge slightly, leaving room for air flow. A filtration fabric placed in the side vents allows air to move through while preventing insects from entering.

Right triangle A three sided shape that includes a 90-degree angle between two of its sides.

Rim joist A joist that runs around the perimeter of the floor joists and home.

Rimless sink a sink with edges that overlap the hole in the countertop. Rimless sinks are usually made of heavy materials such as cast iron.

Rimmed sink A sink with a rim that attaches to the edge of the sink and to the countertop.

Rise The vertical distance from the eaves line to the ridge.

Rise The vertical distance from stair tread to stair tread.

Riser Each of the vertical boards closing the spaces between the treads of stairways.

Riser and panel The exterior vertical pipe (riser) and metal electric box (panel) the electrician provides and installs at the rough electric stage.

Road base An aggregate mixture of sand and stone.

Rock 1, 2, 3 When referring to drywall, this means to install drywall to the walls and ceilings (with nails and screws) before taping is performed.

Rodding Tamping technique that involves consolidating concrete by the use of a push stick or rod.

Roll (rolling) To install the floor joists or trusses in their correct place.

(To "roll the floor" means to install the floor joists.)

Rolled wall covering Any wall covering that is provided in rolls. Examples include fabric, vinyl, and paper.

Roll roofing A roofing material produced in rolls, made by saturating organic mat with asphalt or coal tar pitch and embedding mineral granules on the surface exposed to the weather. Rolls are usually 36 in. wide and 108 sq. ft. of material. Weights are generally 45 to 90 pounds per roll.

Roll tiles Tile shingles that use caps and pans that form a series of peaks and valleys on the finished roof. Roll tiles include barrel and S-tiles.

Roll type standing seam Standing seam roof in which the panels are placed next to each other with standing edges touching. The edges are then mechanically crimped to fasten and seal the seam.

Romex A named brand of nonmetallic sheathed electrical cable that is used for indoor wiring.

Roof deck Surface of the sheathing placed over the roof framing.

Roof diaphragm The entire roof system including rafters or trusses, bracing, sheathing, rough fascia, ridge boards, fasteners, and so forth. All elements of the roof system work together to form a diaphragm that resists wind and other forces and secure the top of exterior walls.

Roof drain Used with a roof membrane system, it fastens into the roof deck and carries water into a drainpipe. It is usually covered with a strainer that filters out leaves and other debris that may clog the drainpipe. You should consider other options before installing a roof drain as they do clog up, especially in cold climates.

Roofing felt An asphalt-saturated organic mat that is produced in rolls. Used as shingle or siding underlayment or anywhere a moisture-resistant barrier is needed; also called tar paper or organic felt.

Roof jack Sleeves that fit around the black plumbing waste vent pipes at, and are nailed to, the roof sheeting.

Roof joist The rafters of a flat roof; lumber used to support the roof sheeting and roof loads. Generally 2 in. × 10 in. and 2 in. × 12 in. rafters are used.

Roof sheathing or sheeting The wood panels or sheet material fastened to the roof rafters or trusses on which the shingle or other roof covering is laid.

Roof system This includes the roof framing, sheathing, trusses, and roofing material. It is a structural part because it helps hold the bearing walls in place, resisting forces that attempt to move the walls such as wind and earthquakes.

Roof valley The "V" created where two sloping roofs meet.

Rosette A circular or oval decorative wood piece used at the termination of a stair rail into a wall. See **balustrade**.

Rough electrical, rough-in electrical Any electrical device or part that will be hidden by, or embedded in, the finish wall.

Rough fascia A horizontal member that is fastened to the vertical edge of the rafter tail or truss and later covered by the fascia; also commonly referred to as the subfascia.

Roughing-in The initial stage of a plumbing, electrical, heating, carpentry, or other project, when all components that won't be seen after the second finishing phase are assembled. Completion of the rough-in usually means all parts that penetrate through the wall, floor, and roof sheathing are in place. The rough-in work for these four trades is reviewed in the four-way inspection before the walls or ceilings are covered. See also **electrical rough**, **heat rough**, and **plumbing rough**.

Rough level The initial process of placing wet concrete at the approximate desired level. The final level is applied later, when the surface is finished with the desired texture, after the concrete has lost some, but not all, of its plasticity.

Rough opening The horizontal and vertical measurement of a window or door opening before drywall or siding is installed.

Rough plumbing, rough-in plumbing Generally roughing-in means to bore holes through the studs for the pipes and to install and connect pipes, but it does not include connecting fixtures or any end elements.

Rough sill The framing member at the bottom of a rough opening for a window. It is attached to the cripple studs below the rough opening.

Rounding decimals The process of rounding, either up or down, the numbers after the decimal point to the desired precision (i.e., if the third digit after the decimal point is five or higher and the desired precision is 1/100, you should increase the second digit after the decimal point by one and if the third digit after the decimal point is four or lower, leave the second digit as is).

Rounding dimensions The process of rounding dimensions to the nearest unit of desired precision (i.e., a measurement ending with a fraction less than ⅛ in. and a desired precision of ½ in. should be rounded down to the next lower in.).

Rounding to a unit Rounding up a material quantity calculation so that a fractional portion is equal to a multiple of the amount contained in the smallest package in which the material may be purchased.

Router Power tool used to cut holes or openings into wood panels without the need to start at an edge. A router is also used to make decorative pattern cuts in wood.

Rowlock A course of brick laid on edge with their ends exposed.

Rubberized asphalt membrane A shingle underlayment that adheres to the roof deck and seals around shingle nails driven through it during installation. Also referred to as bituthene, ice shield, or storm shield, it is placed on the roof where ice damming may occur to prevent water that may pass through the shingles from damaging the structure.

Run (roof) The horizontal distance from the eaves to a point directly under the ridge; one-half the span.

Run (stair) The horizontal distance of a stair tread from the nose to the riser.

Rung A rod or bar that forms the step in a ladder. Rungs attach to the two side rails of the ladder.

R-value A measure of insulation; a measure of a material's resistance to the passage of heat. The higher the R-value, the more insulating "power" it has. For example, typical walls in new homes are usually insulated with 4 in. of batt insulation with an R-value of R-13 and a ceiling insulation of R-30.

Sack mix The amount of Portland cement in a cu. yd. of concrete mix.

Saddle A small second roof built behind the back side of a fireplace chimney to divert water around the chimney; also, the plate at the bottom of some, usually exterior, door openings; sometimes called a threshold.

Saddle T A connection that is used to tap into existing water supply lines. The Saddle T is clamped onto the pipe. When the valve is opened a drill bit–like point pierces the pipe and allows water into the Saddle T and the pipe connected to it.

Sailor course A course of brick with each brick set vertically with the face, the long-wide side, of the brick exposed.

Sales contract A contract between a buyer and seller that should explain the following:

1. What the purchase includes
2. What guarantees there are
3. When the buyer can move in

4. What the closing costs are

5. What recourse the parties have if the contract is not fulfilled or if the buyer cannot get a mortgage commitment at the agreed-upon time

Sand float finish Lime that is mixed with sand, resulting in a textured finish on a wall.

Sanitary sewer A sewer system designed for the collection of wastewater from the bathroom, kitchen, and laundry drains; usually not designed to handle stormwater.

Sapwood Wood found near the surface of the tree, between the bark and the heartwood. Sapwood is lighter in color and less resistant to decay than heartwood.

Sash A single light frame containing one or more lights of glass; also, the frame that holds the glass in a window—often movable part of the window.

Sash A device, usually operated by a spring, designed to hold a single-hung window vent up and in place.

Saturated felt A felt that is impregnated with tar or asphalt.

Saw kerf Groove cut by a circular saw blade.

Saw kerf counterflashing A specially shaped counterflashing that is pressed into a saw kerf cut in masonry to prevent water penetration.

SBS (styrene butadine styrene) A plasticizer used in the hot-mop type of modified bitumen roof systems.

Schedule (windows, doors, hardware, mirrors, etc.) A table on the blueprints that list the sizes, quantities, and locations of the windows, doors, and mirrors.

Scissor truss A truss where the bottom chord is not horizontal. It is used where a sloped ceiling is desired in the inside of the building. The slope of the bottom chord is always less than the slope of the top chord.

Scoring Process of cutting grooves into the face of panels thereby creating a different geometric visual with decorative and sometimes acoustical benefit.

Scrap out The removal of all drywall material and debris after the home is "hung out" (installed) with drywall.

Scratch coat The first coat of plaster, which is scratched to form a bond for a second coat.

Screed (concrete) To level off concrete to the correct elevation during a concrete pour.

Screed (plaster) A small strip of wood, usually the thickness of the plaster coat, used as a guide for plastering.

Scribing Cutting and fitting woodwork to an irregular surface.

Sculptured carpet A design of carpeting that is characterized by a mixture of high-pile and low-pile fibers arranged according to a specific configuration.

Scupper An opening for drainage in a wall, curb, or parapet, usually connected to the downspout.

Sealable equipment Equipment enclosed in a case or cabinet that is provided with a means of sealing or locking so the live parts cannot be made accessible without opening the enclosure. The equipment may or may not be operable without opening the enclosure.

Sealant strip Strips of asphalt placed on the face of the shingle where they will be covered by shingles in the course above. When the shingle is warmed by the sun, the sealant strip adheres to the shingle above, thus creating a tight, wind-resistant connection.

Sealer A finishing material, either clear or pigmented, that is usually applied directly over raw wood for the purpose of sealing the wood surface.

Sealer coat A coat of sealer applied to a substrate to protect it from being stained by something that may drop or spill on it.

Seal tab See **sealant strip**.

Seamless gutter An aluminum gutter is often called a seamless gutter because each straight section is made without seams.

Seasoning Drying and removing moisture from green wood.

Section One of the five basic views found on a plan. A section is a view of the building as though it had been sliced through vertically and opened up so you could see what is inside. It may be thought of as a view showing a dissection of the building.

Sedimentation The addition of soil particles to water bodies by natural and human-related activities. Sedimentation often decreases water quality and can accelerate the aging process of lakes, rivers, and streams.

Select and better oak grade A grade of oak strip flooring in which at least 50% of the wood is clear of defects except for a few small bright spots of sap. The other 50% may have pinworm holes, machine defects, and no more than one small tight knot for every 3 linear ft. of wood. See **number one common oak grade** and **number two common oak grade**.

Self-sealing shingles Shingles containing factory-applied strips or spots of self-sealing adhesive.

Self-tapping screw A type of screw, commonly used with light gauge metal, that has a drill bit–style tip that forms its own hole in the metal.

Semigloss paint or enamel A paint or enamel made so that its coating, when dry, has some luster but is not very glossy. Bathrooms and kitchens are often painted semigloss.

Separately derived system A premise's wiring system whose power is derived from a battery; from a solar photovoltaic system; or from a generator, transformer, or converter windings and that has no direct electrical connection, including a solidly connected grounded circuit conductor, to supply conductors originating in another system.

Septic system An on-site wastewater treatment system. It usually has a septic tank that promotes the biological digestion of the waste and a drain field that is designed to let the leftover liquid soak into the ground.

Serpentine seam (carpet) A carpet seam that is made by seaming two pieces of carpet that have been cut in meandering curved or S-shaped patterns. Serpentine seams are more time consuming and difficult to install than straight seams but are believed to make the seam less visible. See **straight seam**.

Service (electrical) The conductors and equipment for delivering electric energy from the serving utility to the wiring system of the premises served.

Service cable Service conductors made up in the form of a cable.

Service conductors The conductors from the service point to the disconnecting means.

Service drop The overhead service conductors from the last pole or other aerial support to and including the splices, if any, connecting to the service entrance conductors at the building or other structure.

Service entrance conductors (overhead system) The service conductors between the terminal of the service equipment and a point usually outside the building, clear of building walls, where they are joined by tap or splice to the service.

Service entrance conductors (underground system) The service conductors between the terminals of the service equipment and the point of connection to the service lateral.

Service entrance panel The main power cabinet where electricity enters a home wiring system.

Service equipment The main control gear at the service entrance, such as circuit breakers, switches, and fuses.

Service lateral The underground service conductors between the street main, including any risers at a pole or other structure or from transformers, and the first point of connection to the service entrance conductors in a terminal box or meter or other enclosure, inside or outside the building wall. Where there is no terminal box, meter, or other enclosure, the point of connection is considered to be the point of entrance of the service conductors into the building.

Service point The point of connection between the facilities of the serving utility and premise's wiring.

Setback The distance from the property line to the foundation of the structure. Minimum setbacks are established by the local government to maintain desired appearance standards by keeping structures from being built too close to the edge of the property.

Setback thermostat A thermostat with a clock that can be programmed to come on and off at various temperatures and at different times of the day or week.

Settlement Shifts in a structure, usually caused by freeze-thaw cycles underground.

Sewage ejector A pump used to "lift" wastewater to a gravity sanitary sewer line; used in basements and other locations that are situated below the level of the side sewer.

Sewer lateral The portion of the sanitary sewer that connects the interior wastewater lines to the main sewer lines.

Sewer stub/septic tank stub A short section of pipe connected to the main sewer line or septic tank and extending toward the home. It is designed so the drain line coming from the home can easily be connected to it. The footings on a full-basement home should be positioned so that the stub is lower than the bottom of the footings. This ensures a downhill slope for a sewer line extending from under the footing out to the stub.

Sewer tap The physical connection point where the home's sewer line connects to the main municipal sewer line.

Shag carpet A carpet with a long pile. See **berber carpet, indoor-outdoor carpet, nylon carpet, sculptured carpet**, and **wool carpet**.

Shake A wood roofing material, normally cedar or redwood, produced by splitting a block of the wood along the grain line. Modern shakes are sometimes machine sawn on one side. See **shingle**.

Shake felt Roofing felt that usually comes in rolls 18 in. wide.

Shake roof A roof constructed from roofing material made from hand-split wood. Shakes come in three thicknesses—thin, medium, and heavy—and are usually made from cedar with relatively straight grain and free of knots. See also **hand-split and resawn shake** and **tapersawn shake**.

Shaper A machine with revolving cutters that is used to cut moldings and other irregular outlines.

Shear block Plywood that is face-nailed to short 2 in. × 4 in. or 2 in. × 6 in. wall studs (e.g., above a door or window). This is done to prevent the wall from sliding and collapsing.

Shear force Pressure required to break the attachment between two members, causing them to slide across each other. For example, if the nail attaching two panels is severed by shear force, the members will slide.

Shear panel Usually a plywood or oriented strand board sheet that covers the wall from the top plate to the bottom plate. When nailed in place, this sheet resists shear forces applied to the wall that try to move it out of square.

Sheathing, sheeting The structural wood panel covering, usually OSB or plywood, used over studs, floor joists, rafters, or trusses of a structure.

Shed roof A roof containing only one sloping plane.

Sheet metal duct work The heating system. In a house, it is usually made of round or rectangular metal pipes and sheet metal and installed for distributing warm (and cold) air from the furnace or air-conditioner to rooms in the home.

Sheet metal work All components of a house employing sheet metal, such as flashing, gutters, and downspouts.

Sheet rock A manufactured panel made out of gypsum plaster and encased in a thin cardboard. "Green board"–type drywall has a greater resistance to moisture than regular (white) plasterboard and is used in bathrooms and other "wet areas."

Sheet siding Exterior finish material that comes in sheets, usually 4 ft. wide by 8, 9, or 10 ft. long.

Sheet stock Materials such as plywood and medium-density fiberboard that comes in sheets.

Shim A small piece of scrap lumber or shingle, usually wedge shaped, that when forced behind a furring strip or framing member, forces it into position. Also used when installing doors and placed between the door jamb legs and 2 in. × 4 in.

door trimmers. Metal shims are wafer 1½ in. × 2 in. sheet metal of various thickness used to fill gaps in wood framing members, especially at bearing point locations.

Shingles Roof covering of asphalt, wood, tile, slate, or other material cut to stock lengths, widths, and thicknesses.

Shingles (siding) Various kinds of shingles, used over sheathing for exterior wall covering of a structure.

Shingle tack coat On composition shingles, the shingle mat after it has been saturated with asphalt or coal tar pitch but before granules, talc, or other materials have been embedded into the surface.

Shiplap siding Horizontal siding that has been rabbeted on both long edges. A weathertight connection is formed when the rabbet joint on the upper piece overlaps the rabbet joint on the bottom piece.

Short circuit A situation that occurs when hot and neutral wires come in contact with each other. Fuses and circuit breakers protect against fire that could result from a short.

Shovel footing A footing form that is typically made by thickening the concrete floor slab, usually formed by using a shovel to trench the area that is to be filled by the shovel footing.

Shower pan Noncorrosive pan that covers the base of the shower and runs partway up the wall.

Shutter Louvered decorative frames in the form of doors located on the sides of a window. Some shutters are made to close over the window for protection.

Side jamb A specific name for the jamb located on each side of the inside of window and door openings.

Side sewer The portion of the sanitary sewer that connects the interior wastewater lines to the main sewer lines. The side sewer is usually buried in several ft. of soil and runs from the house to the sewer line. It is usually "owned" by the sewer utility, must be maintained by the owner, and may only be serviced by utility approved contractors; sometimes called sewer lateral.

Sidewalk, lighting Paved or otherwise improved areas for pedestrian use located within public street rights-of-way also containing roadways for vehicular traffic.

Sidewall pressure The force exerted on a cable as it is dragged around a bend. The longer the pull and the tighter the bend radius, the higher the sidewall pressure will become. High sidewall pressure damages cable.

Siding The finished exterior cover-ing of the outside walls of a frame building.

Siding (lap siding) Slightly wedge-shaped boards used as horizontal siding in a lapped pattern over the exterior sheathing; varies in butt thickness from ½ to ¾ in. and in widths up to 12 in.

Siding batten Long, narrow strip of trim commonly used to cover vertical joints on vertical exterior siding.

Silicon A chemical element—Si, atomic number 14, semimetallic in nature, and dark gray—that is an excellent semiconducting material and is the most common semicon-ducting material used in making pho-tovoltaic devices.

Sill The 2 in. × 4 in. or 2 in. × 6 in. wood plate framing member that lays flat against, and is bolted to, the foundation wall (with anchor bolts) and upon which the floor joists are installed. Normally the sill plate is treated lumber.

Sill The member forming the lower side of an opening, as a door sill or window sill.

Sill cock An exterior water faucet (hose bib).

Sill plate (mudsill) The bottom horizontal member of an exterior wall frame that rests on top a foun-dation; sometimes called mudsill; also sole plate, bottom member of an interior wall frame.

Sill seal Fiberglass or foam insula-tion installed between the foundation wall and sill (wood) plate; designed to seal any cracks or gaps.

Silt barrier Material placed over the course aggregate of a perimeter drain system that allows water to enter the drain system while preventing silt (i.e., dirt) from filtering down and clogging the system.

Simple payback The amount of time it will take to recover the initial investment through savings. The sim-ple payback (in years) can be calcu-lated by dividing first cost by annual savings. The simple payback method does not consider the time value (interest) of money. Therefore its results should be considered rough, at best.

Simplex communications sys-tem A communications system in which data can only travel in one direction.

Single-element transducer A transducer having one measuring element.

Single-hung window A window with one vertically sliding sash or window vent.

Single membrane A roof sys-tem with just one waterproof layer. The most common types of single-membrane roofs are modified bitumen and elastomeric roof systems.

Single-pane glass A window pane that has only one sheet of glass. See **thermal pane** and **triple pane**.

Single phase Single-phase electric power refers to the distribution of electric power using a system in which the voltage is taken from one phase of a three-phase source. Single-phase distribution is used when loads are mostly lighting and heating, with few large elements.

Single phase This implies a power supply or a load that uses only two wires for power. Some "grounded," single-phase devices also have a third wire used only for a safety ground, which is not connected to the electrical supply or load in any other way except as a safety ground.

Single-wall flue A flue consisting of a single metal pipe.

Sinkers Teflon-coated common nails used to minimize the splitting of lumber because they are easier to hammer into the wood; disapproved by some engineers because the Teflon coating that allows them to more easily slip into the wood may also allow them to more easily slip out of the wood.

Sintered plate (battery) The plate of an alkaline cell, the support of which is made of sintered metal powder and into which the active material is introduced.

Site area The total project area including all areas of property, constructed and nonconstructed areas.

Six points of estimation A method for estimating a loss using six steps. Combining the first letter of each step spells the word points. The six points of estimation are perspective, organization, identification, number, technique, and supporting events.

Sizing A compound that is placed on wood, plaster, or other porous surfaces to fill the pores thus preparing the surface for additional finishes.

Skin effect In an air-conditioning system, the tendency of the outer portion of a conductor to carry more of the current as the frequency of the air-conditioning increases.

Skirt Decorative trim, usually made from a wood board, that is installed on the wall below exposed stairs, which trims the area around or just below the exposed ends of the treads and risers. See **balustrade** and **stair bracket**.

Skylight A more or less horizontal window located on the roof of a building.

Slab (concrete) Concrete pavement (i.e., driveways, garages, and basement floors).

Slab (door) A rectangular door without hinges or frame.

Slab cabinet door A cabinet door made from a single piece of material. A flush slab door is typically made from medium density fiberboard (MDF) or plywood and is either painted or covered with veneer. Sometimes, decorative patterns are carved into its surface or decorative moldings are attached. See **frame and panel cabinet door**.

Slab on grade A type of foundation with a concrete floor that is placed directly on the soil. The edge of the slab is usually thicker and acts as the footing for the walls.

Slag Concrete cement that sometimes covers the vertical face of the foundation void material.

Slate Heavy metamorphic rock available in several different colors; used in flooring, roofing, and wall panels. Roofing slate comes in a variety of colors classified as unfading or weathering. Unfading colors stay very close to their original color throughout their life. Weathering colors change as they age.

Sleeper Usually, a wood member embedded in concrete, as in a floor, that serves to support and to fasten the subfloor or flooring.

Sleeve(s) Pipe installed under the concrete driveway or sidewalk and will be used later to run sprinkler pipe or low-voltage wire.

Sliding T-bevel A hand tool with an edge that can be adjusted and then locked into position to mark angles for specific layouts.

Sliding window A window unit that opens by sliding one window sash past another horizontally.

Slip-matched veneer Veneer produced by sliding or slipping pieces of veneer next to each other. The grain of slip-matched veneer appears to run along the entire surface. See **book-matched veneer, veneer, unmatched veneer,** and **whole piece veneer**.

Slip sheet Light roofing paper or thin fabric that allows the PVC roof membrane to easily slip over the foam insulation without rubbing and suffering damage.

Slope The incline angle of a roof surface, given as a ratio of the rise (in in.) to the run (in ft.). See also **pitch**.

Slump The "wetness" of concrete. A 3-in. slump is dryer and stiffer than a 5-in. slump.

Slump block A masonry unit that is made by removing the forms before the concrete is completely dry. The concrete sags, or slumps, causing the block to have a rounded look. Slump block may be colored with a concrete dye admixture or by painting the surface of the block; also called slump stone.

Smoke shelf Ledge in the masonry flue that prevents downdrafts and moisture from entering the firebox.

Smoothwall texture A drywall finish with no visible texture. To prevent flashing, the entire surface is coated with a thin surface coat.

Snap-type standing seam A standing seam roof in which the cap edge is snapped into place over the underlying edge to lock the edges in place and provide a watertight seal.

Soap Slang for cable-pulling lubricant.

Soffit The area below the eaves and overhangs. The underside where the roof overhangs the walls; usually the underside of an overhanging cornice.

Soil pipe A large pipe that carries liquid and solid wastes to a sewer or septic tank.

Soils engineer A licensed engineer who performs the necessary calculations to determine the types and sizes of footings, retaining walls, and the like; also called a geotechnical engineer.

Soil stack A plumbing vent pipe that penetrates the roof.

Solar cell Photovoltaic cell.

Solar energy Energy from the sun. The heat that builds up on surfaces exposed to the sun is an example.

Soldier course A course of brick with each brick set vertically with the edge, the long-narrow side, of the brick exposed.

Sole plate The bottom, horizontal framing member of a wall that's attached to the floor sheeting and vertical wall studs.

Solid bridging A solid member placed between adjacent floor joists near the center of the span to prevent joists or rafters from twisting.

Solid plastic countertop A class of countertops made from plastic resins; includes cultured countertops and solid surface countertops. Solid plastic materials are also used to make tub and shower surrounds.

Solid surface countertop A countertop made from plastic resin. Solid surface materials are also used to make tub and shower surrounds.

Sonotube A round, large cardboard tube designed to hold wet concrete in place until it hardens.

Soot mapping A phenomenon that occurs when soot collects on a wall in a way that reveals or maps materials that are hidden in the wall finish such as drywall tape, the edges of drywall boards, and screws or nails.

Sound attenuation soundproofing a wall or subfloor, generally with fiberglass insulation.

Spaced sheathing A sheathing material that is installed to allow air to flow between it and in and around wood shingles installed on it. Spaced sheathing is used because it helps wood shingles last longer by keeping them uniformly dry.

Space heat Heat supplied to the living space (e.g., to a room or the living area of a building).

Spacing The distance between individual members or shingles in building construction.

Spalling A condition where the surface of the concrete flakes off. It can be caused by premature trowling, overworking the concrete, exposure to high heat or chemicals, or water penetrating the surface and freezing.

Span The clear distance that a framing member carries a load without support between structural supports; also, the horizontal distance from eaves to eaves and the distance between two poles of a transmission or distribution line.

Spark test A high-voltage test performed on certain types of conductor during manufacture to ensure the insulation is free from defects.

Spill light Unwanted light directed onto a neighboring property; also referred to as light trespass.

Spline A strip of metal or fiber inserted in the kerfs of adjacent acoustical tile to form a concealed mechanical joint seal.

Split phase A split-phase electric distribution system is a three-wire, single-phase distribution system commonly used in North America for single-family residential and light commercial (up to about 100 kVA) applications.

Square A situation that exists when two elements are at right angles to each other; also, a tool for checking this.

Square foot A two-dimensional calculation (e.g., the area of a square).

Square-tab shingles Shingles on which tabs are all the same size and exposure.

Square yard A two-dimensional calculation (e.g., the area of a square). There are 9 sq. ft. in a sq. yd. (i.e., 3 ft. × 3 ft.).

Squaring a wall Pulling the corners of the wall so that the diagonal distance from corner to corner is equal, which means that the wall section forms a perfect rectangle. A wall is held in this shape by let-in bracing or shear panels.

Squeegie Fine pea gravel used to grade a floor (normally before concrete is placed).

Stack (trusses) To position trusses on the walls in their correct location.

Stack bond A course of brick in which each brick is directly over the brick in the course below it, making all the vertical joints form a line.

Stage A subsection of a project or a group of tasks that are performed together and that have specified and scheduled outcomes.

Stain-grade material Trim material that has few flaws and is suitable for use in materials that will be stained, leaving the grain exposed.

Stair bracket Decorative trim that is attached to the wall or skirt below each stair tread. See **balustrade** and **skirt**.

Stair carriage or stringer Supporting member for stair treads; usually a 2 in. × 12 in. plank notched to receive the treads; sometimes called a rough horse.

Stair clamps After the framer has determined the riser height and tread width, he or she marks these dimensions on the framing square by attaching stair clamps to each leg of the square. The framing square with the

stair clamps attached is used to lay out the stringer. The framer places the framing square on the stringer board until the clamps touch the board and then traces the square.

Stair landing A platform between flights of stairs or at the termination of a flight of stairs; often used when stairs change direction.

Stair rise The vertical distance from stair tread to stair tread.

Stair-step pattern Reference to the installation of several courses of shingles simultaneously with the lowest or bottom course extending further than the next course up and so forth. The result is a zigzag or stair-step outline.

Standard baseboard and casing Standard baseboard is usually 2¼ or 3¼ in. high. Standard casing is usually 2¼ in. wide.

Standard cabinet door hinge A cabinet hinge that is attached to the door on one side and to the cabinet stile on the other side. Standard cabinet door hinges usually cannot be adjusted once they are installed. See **European-style cabinet door hinge**.

Standard operating procedure(s) (SOP) Detailed, written instructions documenting a method to achieve uniformity of performance.

Standard practices of the trade(s)
One of the more common basic and minimum construction standards. This is another way of saying that the work should be done in the way it is normally done by the average professional in the field.

Standing seam Metal roof seam made by turning the long edges of the panels up and then over. The three common types of standing seams are rolled type, snap type, and batten type.

Starter course First row of shingles laid at the eave line. The starter course for composition shingles usually consists of shingles that are installed wrongside down or are made from rolled starter strip material. The starter course for wood shingles or shakes is usually made by sawing 2 to 3 in. off the length and installing the wood shingles or shakes rightside down. The first course of wood shingles or shakes completely overlaps the starter course. The starter course for tiles is also the first.

Starter strip Asphalt roofing applied at the eaves that provides protection by filling in the spaces under the cutouts and joints of the first course of shingles.

Static vent A vent that does not include a fan.

STC (sound transmission class)
The measure of sound stopping of ordinary noise.

Steel beam Two common types are the wide flange steel beams that look like the letter "H" laid sideways and the I beams that look like the capital letter "I."

Steel inspection A municipal or engineers inspection of the concrete foundation wall, conducted before concrete is poured into the foundation panels; done to ensure that the rebar (reinforcing bar), rebar nets, void material, beam pocket plates, and basement window bucks are installed and wrapped with rebar and complies with the foundation plan.

Step flashing A flashing application method used where a vertical surface meets a sloping roof plane. 6 in. × 6 in. galvanized metal bent at a 90-degree angle and installed beneath siding and over the top of shingles. Each piece overlaps the one beneath it the entire length of the sloping roof (step by step).

Stere Another name for cubic meter (m^3); seldom used, except on crossword puzzles.

Stick built A house built without prefabricated parts; also called conventional building.

S tile A tile with a serpentine "S" shape; also commonly referred to as Spanish tile.

Stile An upright framing member in a panel door.

Stone countertop A countertop made from stone such as granite or marble. See **cultured countertop, cultured marble countertop, plastic laminate countertop, solid plastic countertop, solid surface countertop, tile countertop**, and **wood block countertop**.

Stool A member that forms the horizontal shelf at the bottom of the window.

Stop box Normally a cast iron pipe with a lid that is placed vertically into the ground, situated near the water tap in the yard, and where a water cutoff valve to the home is located (underground). A long pole with a special end is inserted into the curb stop to turn the water on or off.

Stop order A formal, written notification to a contractor to discontinue some or all work on a project for reasons such as safety violations, defective materials or workmanship, or cancellation of the contract.

Stops Moldings along the inner edges of a door or window frame; also, valves used to shut off water to a fixture.

Stop valve A device installed in a water supply line, usually near a fixture, that permits an individual to shut off the water supply to one fixture without interrupting service to the rest of the system.

Storm sash or storm window An extra window usually placed outside of an existing one, as additional protection against cold weather.

Storm sewer A sewer system designed to collect stormwater that is separated from the wastewater system. In the past, cellar drains were often connected to the storm sewer system. Nowadays, these drains are usually required to tie into the sanitary sewer system.

Stormwater runoff Water from precipitation that flows over surfaces into sewer systems or bodies of water. All precipitation that leaves project site boundaries on the surface is considered stormwater runoff.

Storm window Additional window unit, complete with a window pane installed in a window sash, installed over the original window unit to provide an extra layer of glass insulation.

Story That part of a building between any floor or between the floor and roof.

Story pole A pole with lines on its surface to mark the height for each row of brick being laid.

Straight seam A carpet seam that is made by seaming two pieces of carpet that have been cut in a straight line. See **serpentine seam**.

Straight wood floor installation An installation where wood strips are installed in straight rows that are usually parallel to at least one of the walls. See **diagonal wood floor installation** and **herringbone wood floor installation**.

Strike The plate on a doorframe that engages a latch or dead bolt.

String, stringer A timber or other support for cross members in floors or ceilings. In stairs, the supporting member for stair treads. Usually a 2 in. × 12 in. plank notched to receive the treads.

Strip flooring Wood flooring consisting of narrow, matched strips.

Strippable wallpaper Wallpaper with a face that easily strips from the backing; also called peelable wallpaper.

Structural damage Damage that affects the ability of a part or parts to hold and carry parts of the structure it was designed to hold and carry; also see **cosmetic damage**.

Structural engineer An engineer licensed to determine the material type, grade, size, and placement requirements for safe construction

of the structural parts used in a building.

Structural floor A framed lumber floor that is installed as a basement floor instead of concrete. This is done on very expansive soils.

Structural part A part of a building that is essential in supporting a load or keeping the structure intact. A structural part cannot be removed without weakening the structure.

Struts Member positioned between two other members to keep them a specific distance apart, giving them added strength.

Stub (stubbed) To push through.

Stubbed out Term used to describe leaving the end of a part exposed for easy connection later in the construction process. Rebar may be stubbed out of the footing for connection to the foundation concrete, or a short section of sewer line may be stubbed out from the septic tank or main sewer line for easy connection to the sewer lateral later in the construction process.

Stucco Refers to an outside plaster finish made with Portland cement as its base.

Stucco sheathing Wall covering on which synthetic stucco is installed. Common materials used are foam board and exterior grade gypsum board.

Stud A vertical wood-framing member, also referred to as a wall stud, attached to the horizontal sole plate below and the top plate above. Normally 2 in. × 4 in. or 2 in. × 6 in. and 8 ft. long (sometimes 92 5/8 in. long); one of a series of wood or metal vertical structural members placed as supporting elements in walls and partitions.

Stud framing A building method that distributes structural loads to each of a series of relatively lightweight studs; contrasts with post-and-beam.

Stud gun Tool that uses gunpowder contained in a cartridge to drive a nail into a hard surface like steel or concrete.

Stud shoe A metal, structural bracket that reinforces a vertical stud; used on an outside bearing wall where holes are drilled to accommodate a plumbing waste line.

Subcontractor A contractor who specializes in performing a specific building trade such as drywall, masonry, or painting. A subcontractor will often enter into a subcontract with a general contractor to perform specific work in the construction for an agreed upon price.

Subfloor The framing components of a floor to include the sill plate, floor joists, and deck sheeting over which a finish floor is to be laid.

Submetering A method of determining the proportion of energy used within a building attributable to specific mechanical end uses or subsystems (i.e., the heating subsystem of an HVAC system).

Substantial completion The stage or designated portion of a construction project that is sufficiently complete in accordance with a contract for the owner to occupy or utilize it for its intended use without undue interference. Payment less the final retention is often based on substantial completion.

Substrate Surface or support onto which a finish surface is placed.

Sump A pit or large plastic bucket or barrel inside the home designed to collect ground water from a perimeter drain system.

Sump pump A submersible pump in a sump pit that pumps any excess ground water to the outside of the home.

Supporting events The sixth of the six points of estimation of a loss. Supporting events is where the estimator includes work that must be done on undamaged items that are required in order to fix damaged items. For example, if a countertop must be replaced, an undamaged sink must be detached and stored until the countertop has been replaced then reset. Because it was not damaged, detaching and resetting the sink is a supporting event to the countertop replacement.

Surface drying A drying process where lumber is allowed to remain exposed to the air long enough to allow it to lose its excess moisture. Lumber dried in this way is marked with an S-Dry stamp.

Surface-mount medicine cabinet Medicine cabinet that is attached to the surface of the wall. See **recess mount medicine cabinet**.

Surfacing Name of the process when rough-cut lumber is planed down to make the surfaces smooth. Sharp knives are run over the surface of the lumber cutting away ¼ in. of the board. When lumber is surfaced on two sides it is called S-2-S, and when it is surfaced on all 4 sides, it is called S-4-S. Framing lumber is generally S-4-S. The resulting dimensions of the board are called nominal dimensions.

Surge withstand A measure of an electrical device's ability to withstand high-voltage or high-frequency transients of short duration without damage.

Suspended ceiling A ceiling system supported by hanging it from the overhead structural framing.

Suspension system (ceiling) A metal grid suspended from hanger rods or wires, consisting of main beams and cross tees, clips, splines, and other hardware that supports lay-in acoustical panels or tiles. The completed ceiling forms a barrier to sound, heat, and fire. It also absorbs in-room sound and hides ductwork and wiring in the plenum.

Sway brace Metal straps or wood blocks installed diagonally on the inside of a wall from bottom to top plate, to prevent the wall from twisting, racking, or falling over in a "domino" fashion.

Switch A device that completes or disconnects an electrical circuit.

Switch (bypass isolation) A manually operated device used in conjunction with a transfer switch to provide a means of directly connecting load conductors to a power source and of disconnecting the transfer switch.

Switch (general-use) A switch intended for use in general distribution and branch circuits. It is rated in amperes, and it is capable of interrupting its rated current at its rated voltage.

Switch (general-use snap) A form of general-use switch constructed so that it can be installed in device boxes or on box covers or otherwise used in conjunction with wiring systems recognized by the NEC.

Switch (isolating) A switch intended for isolating an electric circuit from the source of power. It has no interrupting rating, and it is intended to be operated only after the circuit has been opened by some other means.

Switch (motor-circuit) A switch rated in horsepower that is capable of interrupting the maximum operating overload current of a motor of the same horsepower rating as the switch at the rated voltage.

Switch (transfer) An automatic or nonautomatic device for transferring one or more load conductor connections from one power source to another.

Switchboard A large single panel, frame, or assembly of panels on which are mounted switches, over-current and protection devices, buses, and usually, instruments on the face, back, or both. Switchboards are generally accessible from the rear as well as from the front and are not intended to be installed in cabinets.

Symbols Found on plans, symbols are used to represent common objects such as doors and light switches.

Synthetic felt pad A carpet pad made from man-made felt that is highly resistant to tearing. See **high-density urethane foam pad**, **rebond pad**, and **waffle-type sponge rubber pad**.

Synthetic stucco Stucco that comes premixed by the manufacturer. It is usually applied in two coats that are much thinner than common stucco. It is applied on stucco sheathing.

T & G (tongue and groove) A joint made by a tongue (a rib on one edge of a board) that fits into a corresponding groove in the edge of another board to make a tight flush joint. Typically, the subfloor plywood is T & G.

T & P valve Valve that releases water pressure when temperature and pressure exceed a preset limit.

Tab The exposed portion of strip shingles defined by cutouts.

Tabbed shingle Common type of composition shingle. A tabbed shingle has two to six tabs, but three is the most common number of tabs. Tabbed shingles may have an imprinted texture on their surface.

Tail beam A relatively short beam or joist supported in a wall on one end and by a header at the other.

Takeoff An estimator's estimate of the materials necessary to complete a job.

Tamping The process of pressing plastic material into a confined space using a bar or rod so that it compacts the material, removes air pockets, and causes it to mold completely to the shape of the space into which it is being pressed. Tamping concrete causes the concrete to flow around rebar and under and around window bucks while removing the air pockets that cause honeycombing.

Tapersawn shake Wood shake that is resawn on both faces.

Taper siding Siding with one edge much wider than the other. The thicker edge may have a groove or rabbet cut out of it so that it fits snugly over the thin edge of the course of siding directly below it.

Taping The process of covering drywall joints with paper tape and joint compound.

Technique The fifth of the six points of estimation of loss. Technique is where the estimator decides how a damaged item will be replaced. For example, technique is where an estimator decides whether an item should be repaired, replaced, cleaned, or painted.

Teco Metal straps that are nailed and secure the roof rafters and trusses to the top horizontal wall plate; sometimes called a hurricane clip.

Tee-A T-shaped plumbing fitting.

Teflon tape Tape made from Teflon that is wrapped around threads and helps to seal threaded pipe joints.

Tegular tile Ceiling tiles with recessed edges that allow the tile to hang below the ceiling grid.

Tempered glass Tempered glass will not shatter or create shards but will "pelletize." Most codes require tempered glass in tub and shower enclosures, entry doors and sidelights, and in a window when the window sill is less than 16 in. to the floor. Tempered glass is expensive.

Tenon A projection formed on the end of a timber or the like for insertion into a mortise of the same dimensions.

Tension A pulling or stretching force. Tension is the opposite of compression.

Termites Wood-eating insects that superficially resemble ants in size and general appearance and live in colonies.

Termite shield A shield, usually of galvanized metal, placed in or on a foundation wall or around pipes to prevent the passage of termites.

Terne Material used to make metal roofing panels. Terne is made from steel mixed with 2% copper.

Terra cotta A ceramic material molded into masonry units.

Terrazzo Traditionally a type of stone flooring made from marble or other stone chips that are mixed in Portland cement, poured in place, allowed to dry, and then polished. In the 1970s, polymer-based terrazzo was introduced and is called thin-set terrazzo. Today, most of the terrazzo installed is epoxy terrazzo.

Thermal break Insulating layer located between the inside and outside parts of an aluminum window frame to block the flow of heat through the window frame.

Thermal expansion The expansion of a material when subjected to heat.

Thermally protected (as applied to motors) The words "thermally protected" appearing on the nameplate of a motor or motor-compressor indicate that the motor is provided with a thermal protector.

Thermal pane glass A window pane with two sheets of glass and a spacer between them. See **single-pane glass** and **triple-pane glass**.

Thermal protector (as applied to motors) A protective device for assembly as an integral part of a motor or motor-compressor that, when properly applied, protects the motor against dangerous overheating due to overload and failure to start. The thermal protector may consist of one or more sensing elements.

Thermoplastic A plastic compound that will soften and melt with sufficient heat. Thermoplastic insulation compounds are used to manufacture certain types of electrical cables.

Thermoply An exterior laminated sheathing nailed to the exterior side of the exterior walls. Normally ¼-in. thick, 4 ft. × 8 ft. or 4 ft. × 10 ft. sheets with an aluminumized surface.

Thermoset A plastic compound that will not remelt. Thermoset insulation compounds are used to manufacture certain types of cables.

Thermostat A device that relegates the temperature of a room or building by switching heating or cooling equipment on or off.

THHN A thermoplastic-insulated, nylon-jacketed conductor designed for use in dry locations and in operating temperature of up to 90°C.

Thin-set tile Tiles that are attached directly to a substrate such as drywall. See **isolation membrane**.

Three-dimensional calculations A process by which the number of three-dimensional units (e.g., cu. yd.) is determined for a given structural part. Three-dimensional units of measure include all those that are measured in three directions (e.g., length, width, and height). Three-dimensional units of measure deal with volume. Examples include cu. ft., cu. yd., and board ft.

Three-dimensional shingles Laminated shingles. Shingles that have added dimensionality because of extra layers or tabs, giving a shakelike appearance; may also be called architectural shingles.

Three-way switch Electrical switches used to control the same fixture from two different locations, such as two ends of a hall.

Threshold The bottom metal or wood plate of an exterior door frame.

Throat The area at the top of the firebox between the face of the smoke shield and the top of the flue.

Tile base A specialty tile trim piece installed on a wall that covers the corner of a floor and the wall. See **cap mold piece**, **cove piece**, and **double bullnose**.

Tile countertop A countertop made from tiles that are glued to a substrate. See **cultured countertop**, **cultured marble countertop**, **plastic laminate countertop**, **solid plastic countertop**, **solid surface countertop**, **stone countertop**, and **wood block countertop**.

Time and materials contract A construction contract that specifies a price for different elements of the work, such as cost per hour of labor, overhead, profit, and so on. A contract that may not have a maximum price or may "state a price not to exceed."

Tin-can stud Slang for a lighter-gauge metal stud.

Tinner A slang name for the heating contractor.

Tip up The downspout extension that directs water (from the home's gutter system) away from the home. They typically swing up when mowing the lawn and so on.

Title Evidence (usually in the form of a certificate or deed) of a person's legal right to ownership of a property.

TJI or TJ A manufactured structural building component resembling the letter "I"; used as floor joists and rafters. I-joists include two key parts: flanges and webs. The flange of the I-joist may be made of laminated veneer lumber or dimensional lumber, usually formed into a 1½-in. width. The web or center of the I-joist is commonly made of plywood or oriented strand board (OSB). Large holes can be cut in the web to accommodate ductwork and plumbing waste lines. I-joists are available in lengths up to 60 in. long.

T-lock shingle Most common type of interlocking shingle. Produces a basket-weave pattern by sliding the lower edge of the shingles into slots at the top of the downhill shingles.

Toe kick Bottom portion of a lower cabinet unit that is recessed to reduce damage from shoes and hide marring that occurs as shoes hit against the finished material.

Toe nailing To drive a nail in at a slant. A method used to secure floor joists to the plate.

Ton (HVAC) In cooling, it is the amount of cooling an air-conditioning unit provides. One ton is equal to 12,000 BTUs.

Ton (weight) Unit of weight equal to 2000 pounds.

T1-11 (Texture 1-11) Sheets of wood siding textured with a series of evenly spaced vertical grooves.

Tongue and groove A type of edge often found on materials to be used for sheathing and flooring. Each panel has one long edge with a tongue and the other long edge with a corresponding groove. The tongue of one sheet will fit into the groove of the next sheet to form a seam or joint.

Tooled joint A mortar joint between courses of brick in which the mason removes the excess mortar so that it is flush with the face of the brick. A tool is then used to shape the mortar.

Tooling The process of removing unwanted material from a finish carpentry joint through the use of a chisel, rasp, or other sharp instrument.

Tool marks Marks left in material by the knives used to create the shape such as those in a shaper or molder.

Top chord The upper or top member of a truss.

Top chord bearing Flat trusses that are hung from their top chord.

Topping mud A type of drywall mud. Topping mud is used for final coats and contains less adhesive chemicals than does all-purpose mud.

Top plate The top horizontal member of a frame wall. The top plate supports ceiling joists, rafters, or other members.

Top rail The top handrail used on a balustrade. The tops of balusters are attached into the underside of the top rail. See **balustrade** and **bread loaf top rail**.

Torch-down installation method An installation method in which roofing materials are heated with a torch until the material liquefies and forms a bond between layers, overlapping seams, or flashing. On modified bitumen roof systems, a method whereby atactic polypropylene–type (APP-type) modified bitumen roofing is adhered to the base sheet.

Townhouse A single-family dwelling unit constructed in a group of three or more attached units. Each unit extends from the foundation to the roof and each unit has open space on at least two sides.

Transformer A device used to change the voltage of an alternating current in one circuit to a different voltage in a second circuit or to partially isolate two circuits from each other.

Transmitter (garage door) The small, push-button device that causes the garage door to open or close.

Trap A plumbing fitting that holds water to prevent air, gas, and vermin from backing up into a fixture.

Trapezoid A four-sided shape with only two parallel sides.

Travelers Two leads that are connected between a three-way switch to allow power to a fixture to be switched on or off from either switch.

Tread The walking surface board in a stairway on which the foot is placed.

Treated lumber A wood product that has been impregnated with chemical pesticides such as CCA (chromated copper arsenate) to reduce damage from wood rot or insects; often used for the portions of a structure that are likely to be in contact with soil and water. Wood may also be treated with a fire retardant.

Tree wire A type of overhead distribution wire that is insulated for momentary contact with tree branches and used as a primary voltage conductor.

Treeing Water treeing is a form of cable insulation degradation where micochannels, which often appear as a tree-like structure in the insulation, develop due to a complex interaction of water, electrical stress, impurities, and imperfections.

Triangle A three-sided shape. When one of the angles is a right angle, equal to 90 degrees, it is called a right triangle.

Trim (exterior) The finish materials on the exterior a building, such as moldings applied around openings (window trim, door trim), siding, windows, exterior doors, attic vents, crawl space vents, shutters, and so on; also, the physical work of installing these materials.

Trim (interior) The finish materials in a building, such as moldings applied around openings (window trim, door trim) or at the floor and ceiling of rooms (baseboard, cornice, and other moldings); also, the physical work of installing interior doors and interior woodwork, including all handrails, guardrails, stairway balustrades, mantles, light boxes, base, door casings, cabinets, countertops, shelves, window sills and aprons, and so on.

Trim (plumbing, heating, and electrical) The work that the "mechanical" contractors perform to finish their respective aspects of work, when the home is nearing completion and occupancy.

Trimmer The vertical stud that supports a header at a door, window, or other opening.

Trimmer bit Type of router bit that contains a roller that guides the blades along a straight edge.

Triple-pane glass A window pane with three sheets of glass for extra insulation. See **single-pane glass** and **thermal-pane glass**.

Trowel A tool with a flat surface used to finish the concrete surface; usually the last tool used to smooth the concrete surface. Its use must be timed carefully; it should be used after the concrete has lost its weep moisture but before it loses all of its plasticity.

Trowel pattern finish Exterior concrete finish that is created by skilled craftsmen when the concrete is ready to be troweled.

Truss (bridge) A truss bridge is a bridge composed of connected elements (typically straight) that may be stressed from tension, compression, or sometimes both in response to dynamic loads. Truss bridges are one of the oldest types of modern bridges.

Truss (house construction) In house construction, an engineered and manufactured roof support member that is essentially a small truss bridge. It is a structural part used to provide the primary support for the floor or roof sheathing. A roof truss system, including the trusses, sheathing, bracing, and fasteners, also provides support for the tops of the exterior bearing walls.

Tube form A cylindrical tube made from compressed and resin-impregnated paper and used to hold wet concrete until it cures; also known by trade names such as sonotube, sleek tube, and smooth tube.

Tub trap Curved, U-shaped section of a bath tub drain pipe that holds a water seal to prevent sewer gasses from entering the home through the tub's water drain.

Tuck carpet installation method A method of installing carpet on a stair in which the carpet is wrapped around the nose of the tread, attached to the riser, and then attached to the inside edge of the tread below. See **waterfall carpet installation method**.

Turbine vent A vent that creates a vacuum in the attic by turning as the warm air escapes thereby pulling more air out.

Turbo toilet A specialty water closet that uses the water pressure from the plumbing lines to force water into the bowl. The turbo toilet uses less water than most other types of water closets. See **low-profile water closet** and **water closet**.

Turnkey A term used when the contractor or subcontractor provides all materials (and labor) for a job.

Turpentine A petroleum-based, volatile oil used as a thinner in paints and as a solvent in varnishes.

Turtle vent A vent positioned several ft. below the ridge. Turtle vents have no moving parts. As the air heats, it becomes less dense and rises through the turtle vent.

Two-dimensional calculations A process by which the number of two-dimensional units (e.g., sq. ft.) is determined for a given structural part. Two-dimensional units of measure include all those that are measured in two directions. Determining the area of a surface that is measured in two-dimensional units, such as sq. yards or sq. ft., is a typical two-dimensional calculation.

2-year, 24-hour design storm The basis of planning stormwater management facilities that can accommodate the largest amount of rainfall expected over a 24-hour period during a 2-year interval.

Type-X drywall Drywall with a gypsum core that contains reinforcing fibers for added fire protection.

U-block Block that looks the same as a standard block from the front or back, but whose cells are open on the top so that grout can flow outward to the other block on each side. The U-block provides for placement of horizontal reinforcing steel and grout to form a bond beam within the course.

Ufer ground A type of ground where the ground wire is connected to the rebar system inside a footing and foundation system; named after Thomas Ufer, the first person to specify it.

UL (Underwriters Laboratories) An independent testing agency that checks electrical devices and other components for possible safety hazards.

Undercoat A coating applied prior to the finishing or top coats of a paint job. It may be the first of two or the second of three coats; sometimes called the prime coat.

Undercoursing shingles Shingles made from low-quality wood with sapwood, flat grain, and possibly loose knots; usually used under higher-quality shingles since it will not waterproof a roof.

Undercover parking Undercover parking is underground, under a deck, under a roof, or under a building; its hardscape surfaces are shaded.

Underground parking Underground parking is a "tuck-under" or stacked structure that reduces the exposed parking surface area.

Underground plumbing The plumbing drains and waste lines that are installed beneath a basement floor.

Underlayment A ¼-in. material placed over the subfloor plywood sheeting and under finish coverings, such as vinyl flooring, to provide a smooth, even surface.

Underlayment A secondary roofing layer that is waterproof or water-resistant installed on the roof deck and beneath shingles or other roof-finishing layer.

Underslab utilities Heating, plumbing, electrical, or other utilities that are placed under the floor slab. They are generally placed in trenches that are then covered with compactable fill before the concrete floor slab is poured.

Union (plumbing) A plumbing fitting that joins pipes end-to-end so they can be dismantled.

Unmatched veneer Strips of veneer placed according to the veneerers' judgment of how the strips look together or with no regard to graining or pattern; also called pleasing-matched veneer. See **book-matched veneer**, **slip-matched veneer**, **veneer**, and **whole piece veneer**.

Untempered hardboard underlayment An underlayment made from hardboard. Only untempered hardboard should be used as vinyl underlayment. See **cement board underlayment**, **gypsum-based underlayment**, **lauan plywood underlayment**, **particleboard underlayment**, **plywood underlayment**, and **underlayment**.

Upflow furnace A furnace that forces air up and out the top of it.

Upper unit (cabinet) Any cabinet unit that is designed to hang on the wall, usually above a lower unit or appliance. See **full-height cabinet**, **lower unit**, and **vanity cabinet**.

Usable wall space Any section of wall along which a piece of furniture or an appliance may be placed. Hallways are generally not considered usable wall space.

Utility easement The area of the earth that has electric, gas, or telephone lines. These areas may be owned by the homeowner, but the utility company has the legal right to enter the area as necessary to repair or service the lines.

Valley The V-shaped area of a roof where two sloping roofs meet. Water drains off the roof at the valleys.

Valley flashing Sheet metal that lays in the V-shaped area of a roof valley.

Valley jack A type of jack rafter that runs from the valley rafter to the ridge board.

Valley rafter A rafter that runs along the valley, forming the valley line.

Valuation An inspection carried out for the benefit of the mortgage lender to ascertain if a property is a good security for a loan.

Valuation fee The fee paid by the prospective borrower for the lender's inspection of the property; normally paid upon loan application.

Vanity cabinet A type of lower-unit cabinet that is designed to hold a bathroom sink. A standard vanity cabinet is slightly shorter than a standard lower unit. See **full-height cabinet**, **lower unit**, and **upper unit**.

Vapor barrier A building product installed on exterior walls and ceilings under the drywall and on the warm side of the insulation. It is used to retard the movement of water vapor into walls and prevent condensation within them. Normally polyethylene plastic sheeting is used.

Variable rate An interest rate that will vary over the term of the loan.

Veneer Thin layer used as a covering to improve the appearance and durability of the product. Brick and stone can be used as veneers covering the exterior of a structure. Wood may also be used as a veneer to cover non-wood surfaces. Veneers are generally nonstructural parts.

Veneer A thin surface layer, usually wood, that is glued to a base made from less expensive materials such as medium density fiberboard (MDF) or plywood. See **book-matched veneer**, **slip-matched veneer**, **unmatched veneer**, and **whole piece veneer**.

Vent A pipe or duct that allows the flow of air and gasses to the outside; also, another word for the moving glass part of a window sash (i.e., window vent).

Vermiculite A mineral used as bulk insulation and also as aggregate in insulating and acoustical plaster and in insulating concrete floors.

Vibration A technique that involves the use of a mechanical device to shake concrete so that it settles tightly around the rebar and window bucks and removes the large air pockets that otherwise cause honeycombing.

Vibration causes the concrete to settle tightly and smoothly against the forms.

View A specific way of looking at a building. There are five basic views on a plan: elevation, floor plan, plot plan, section, and detail.

Vinyl cove Vinyl that wraps a short distance up the wall.

Vinyl-faced wall covering Type of wall-covering with a vinyl face.

Visqueen A 4 mil or 6 mil plastic sheeting.

Volatile organic compounds (VOCs) These compounds are made with organic chemicals. They are released from products that are being used and that are in storage. In sufficient quantities, VOCs can cause eye, nose, and throat irritations, headaches, dizziness, visual disorders, and memory impairment. Some VOCs are known to cause cancer in animals; some are suspected of causing, or are known to cause, cancer in humans.

Voltage A measure of electrical potential. Most homes are wired with 110- and 220-volt lines. The 110-volt power is used for lighting and most of the other circuits. The 220-volt power is usually used for the kitchen range, hot water heater, and dryer.

Voltage to ground For grounded circuits, the voltage between the given conductor and that point or conductor of the circuit that is grounded; for ungrounded circuits, the greatest voltage between the given conductor and any other conductor of the circuit.

Volute A decorative, circular handrail piece used at the bottom of a stair top rail that is installed over a newel post. See **balustrade**, **gooseneck**, **one-quarter turn**, and **top rail**.

Wafer board A manufactured wood panel made out of 1- to 2-in. wood chips and glue; often used as a substitute for plywood in the exterior wall and roof sheathing.

Waffle-type sponge rubber pad A sponge rubber pad that has been shaped to a pattern of alternating bumps and dimples. See **high-density urethane foam pad**, **rebond pad**, and **synthetic felt pad**.

Wainscot A section of wall covering material that starts at the bottom of the wall, then proceeds upward until it is interrupted by a visible border such as a chair rail.

Walers Horizontal bracing, usually two-by-fours, secured to concrete wall forms to stiffen them so they can be more easily straightened. After attaching the walers, the straightening is accomplished by placing a string parallel to the wall form then moving the form into alignment with the string and attaching bracing to hold the wall in position. The walers help hold the areas between the bracing in the straightened position.

Walk-through inspection A final inspection before acceptance of a project. Generally a walk through inspection looks for those last few punch list items that had not yet been accepted by the owner or owner's representative.

Wall out When a painter spray paints the interior of a home.

Wall tie Used at the intersection of two walls to provide a backing for the end stud of the connecting wall. The two types of wall ties are corner ties and wall channels.

Warping Any distortion in a material.

Warranty In construction, there are two general types of warranties. The manufacturer of a product, such as roofing materials or appliances, provides one. The second warranty covers labor. For example, a roofing contract may include a 20-year material warranty and a 5-year labor warranty. Many new homebuilders provide a 1-year warranty. Any major issue found during the first year should be communicated to the builder immediately. Small items can be saved up and presented to the builder for correction periodically through the first year after closing.

Waste Material that must be purchased but cannot be used. Waste can result from trimming, rejection because of defect, or other efforts to maintain acceptable quality of the structural part containing that material. A percent of waste should be included in virtually all material estimates.

Waste disposal Elimination of waste by means of burial in a landfill, combustion in an incinerator, dumping at sea, or any other way that is not recycling or reuse.

Waste diversion Disposal of waste other than through waste disposal as defined in the preceding definition. Examples are reuse and recycling.

Waste pipe and vent Piping that carries wastewater to the municipal sewage system.

Water board Water-resistant drywall that is used in tub and shower locations. Normally green or blue colored.

Water closet Another name for toilet. See **WC**.

Waterfall carpet installation method A method of installing carpet on a stair in which the carpet overlaps the edge of the tread and falls in a straight line to the inside edge of the tread below. See **tuck carpet installation method**.

Water meter pit (or vault) The box or cast-iron bonnet and concrete rings that contains the water meter.

Waterproof Constructed or protected so that exposure to the weather will not interfere with successful operations.

Waterproofing A process of coating the part of the foundation system that will be below the soil level with a material that can withstand long-term exposure to water. Not the same as damp proofing, which can only withstand short-term exposure to water.

Water-repellent preservative A liquid applied to wood to give the wood water-repellant properties.

Water softener A device that removes minerals from water.

Water table The location of the underground water and the vertical distance from the surface of the earth to this underground water.

Water tap The connection point where the home water line connects to the main municipal water system.

Watertight Constructed so that moisture will not enter the enclosure under specific test conditions.

WC An abbreviation for water closet (toilet).

Weather head Device that prevents moisture from entering into the top of the conduit.

Weatherization Work on a building's exterior done to reduce energy consumption for heating or cooling. Work generally includes adding insulation, installing storm windows and doors, caulking cracks, and applying weather stripping; also referred to as winterization.

Weather strip Narrow sections of thin metal or other material installed to prevent the infiltration of air and moisture around windows and doors.

Web Diagonal supporting members running between the top and bottom chords of a truss.

Weep holes Small holes in storm window frames that allow moisture to escape.

Weeping mortar Mortar between courses of brick that has not been trowled or otherwise smoothed after pressing the brick in place.

Weep moisture Excess water contained in the concrete mix that is not needed for hydration. It escapes through all the surfaces of the concrete as the mixture settles and forces it out.

Welded wire mesh (WWM) A grid of heavy gauge wires welded together and used to reinforce concrete slabs; also known as remesh or wire mesh.

Wet-set method A roof tile installation method used on roofs with less than a 7/12 slope and usually over a mineral-faced, hot-mopped underlayment. Mortar is used to hold the tiles in place. The tile is wet before installation so that the mortar will better bond to it. Most commonly used in the Southeastern United States, where high winds and high moisture combine; also referred to as mortar-set method or mud-on method.

Whole house fan A fan designed to move air through and out of a home and normally installed in the ceiling.

Whole piece veneer A piece of veneer that is large enough to cover an entire surface. See **book-matched veneer**, **slip-matched veneer**, and **unmatched veneer**, and **veneer**.

Wind bracing Metal straps or wood blocks installed diagonally on the inside of a wall, from bottom to top plate, to prevent the wall from twisting, racking, or falling over in a "domino" fashion.

Wind locks Metal fastener inserted in the nail hole of a tile shingle and designed to overlap the lip of the next higher tile, providing additional means of holding it in place; also called tile locks.

Window buck Square or rectangular box that is installed within a concrete foundation or block wall. A window will eventually be installed in this "buck" during the siding stage of construction.

Window fin Part of the window unit that serves as the flashing when siding is installed over it.

Window frame The stationary part of a window unit; a window sash fits into the window frame.

Windowpane The glass part of a window unit. Each light in a window unit has a windowpane.

Window sash The operating or movable part of a window. The sash is made of window panes and their border.

Window sill Bottom horizontal member of the window frame.

Window stop A horizontal or vertical piece that prevents the window from falling out of the window frame. The window stop also forms the groove that the window slides in across the surface of the jamb.

Wire gauge A unit of measure used to indicate wire size. The thicker the wire, the smaller the gauge number. See **American standard gauge**.

Wire nut A plastic device used to connect bare wires together.

Wire reinforcement Metal reinforcing mesh placed inside the mortar joints along specified courses or rows of block to tie and reinforce the block. Like the bond beam, courses containing wire reinforcement within the mortar joint tie and reinforce the masonry wall horizontally. A structural engineer specifies which courses should contain the wire reinforcement.

WonderBoard A panel made out of concrete and fiberglass usually used as a ceramic tile backing material; commonly used on bathtub decks.

Wood block countertop A countertop made from solid blocks of wood glued together. See **cultured countertop, cultured marble countertop, plastic laminate countertop, solid plastic countertop, solid surface countertop, stone countertop**, and **tile countertop**.

Wood edge (countertop) An edge on a countertop that is made by trimming the square front corner with a decorative strip of wood. See **plastic laminate countertop**.

Wood shingle roof A roof constructed from wood shingles that are sawn out of logs and are about 3/8-in. thick. Grades of wood shingles are blue label, red label, black label, and undercoursing. They are no longer allowed in fire zones.

Wool carpet A carpet made with natural wool fibers. See **berber**

carpet, indoor-outdoor carpet, nylon carpet, sculptured carpet, and **shag carpet**.

Wrapped drywall Areas that get complete drywall covering, as in the doorway openings of bifold and bypass closet doors.

Wythe Single, vertical masonry wall that is one unit thick. A double-wythe wall is two units thick.

Xeriscaping A landscaping method designed for water conservation so that routine irrigation is not necessary. It includes using drought-adaptable and low-water plants, soil amendments such as compost to conserve moisture, and mulches to reduce evaporation.

Y A Y-shaped plumbing fitting.

Yard of concrete One cu. yd. of concrete is 3 ft. × 3 ft. × 3 ft. in volume or 27 cu. ft. One cu. yd. of concrete will pour 80 sq. ft. of 3½-in. thick sidewalk or basement or garage floor.

Yoke The location where a home's water meter is sometimes installed between two copper pipes; located in the water meter pit in the yard.

Z-bar flashing Bent, galvanized metal flashing that's installed above a horizontal trim board of an exterior window, door, or brick run. It prevents water from getting behind the trim or brick and into the home.

Zero clearance fireplace Another term for factory-built fireplace. The term is misleading. Zero clearance suggests that combustible materials can touch the assembly, when, in reality, no combustible materials should ever touch a factory-built fireplace.

Zone The section of a building that is served by one heating or cooling loop because it has noticeably distinct heating or cooling needs; similarly, the section of property that will be watered from a lawn sprinkler system.

Zone valve A device, usually placed near the heater or cooler, that controls the flow of water or steam to parts of the building; it is controlled by a zone thermostat.

Zoning A governmental process and specification that limits the use of a property (e.g., single-family use, high-rise residential use, industrial use, etc.). Zoning laws may limit where you can locate a structure. Also see **building codes**.

INDEX

Other Manufacturing Titles Certain to Be of Interest

- **Robust Control System Networks: How to Achieve Reliable Control After Stuxnet,** *Ralph Langner*
- **Construction Estimating: A Step-by-Step Guide to a Successful Estimate,** *Karl F. Schmid*
- **ISA-88 and ISA-95 in the Life Science Industries,** *WBF*
- **Applying ISA-88 in Discrete and Continuous Manufacturing,** *WBF*
- **ISA-95 Implementation Experiences,** *WBF*
- **ISA-88 Implementation Experiences,** *WBF*
- **Industrial Resource Utilization and Productivity,** *Edited by Anil Mital, Ph.D. and Arun Pennathur, Ph.D.*
- **Protecting Industrial Control Systems from Electronic Threats,** *Joe Weiss*
- **Advanced Regulatory Control: Applications and Techniques,** *David W. Spitzer*
- **Process Control Case Histories: An Insightful and Humorous Perspective from the Control Room,** *Edited by Gregory K. McMillan*
- **Construction Crew Supervision: 50 Take Charge Leadership Techniques & Light Construction Glossary,** *Karl F. Schmid*
- **Alarm Management for Process Control,** *Douglas H. Rothenberg, Ph.D.*

For more information, please visit **www.momentumpress.net**

The Momentum Press Digital Library

Engineering Ebooks For Research, Classrooms, and Reference

Our books can also be purchased in an e-book collection that features...
- *a one-time purchase; not subscription based,*
- *that is owned forever,*
- *allows for simultaneous readers,*
- *has no restrictions on printing, and*
- *can be downloaded as a pdf*

The **Momentum Press** digital library is an affordable way to give many readers simultaneous access to expert content.

For more information, please visit **www.momentumpress.net/library** *and to set up a trial, please contact info@globalepress.com.*

CPSIA information can be obtained at www.ICGtesting.com
Printed in the USA
BVOW010036140213

313184BV00006B/11/P